Digital Spectral Analysis

C. K. Yuen, D. Fraser

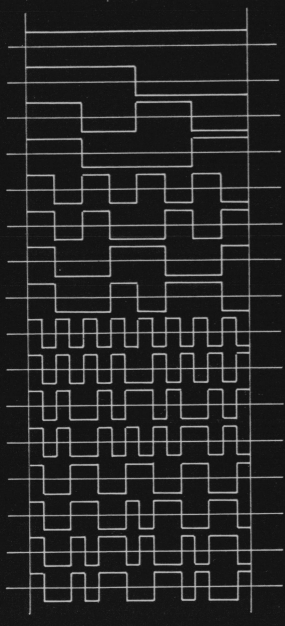

CSIRO / Pitman

Digital Spectral Analysis

Digital Spectral Analysis

C. K. Yuen

Department of Information Science
University of Tasmania

D. Fraser

Division of Computing Research
CSIRO, Canberra

CSIRO

Pitman

National Library of Australia Cataloguing-in-Publication Entry

Yuen, C. K.
  Digital spectral analysis.

  Index
  Simultaneously published, London: Pitman.
  ISBN 0 643 02419 0

  1. Signal processing. 2. Spectral theory (Mathematics).
  I. Fraser, Donald, joint author. II. Commonwealth
  Scientific and Industrial Research Organization.
  III. Title

519.2

Published jointly by

CSIRO
314 Albert Street, East Melbourne, Australia 3002
ISBN 0 643 02419 0

and

PITMAN PUBLISHING LIMITED
39 Parker Street, London WC2B 5PB

FEARON PITMAN PUBLISHERS INC
6 Davis Drive, Belmont, California 94002, USA

Associated Companies
Copp Clark Pitman, Toronto
Pitman Publishing New Zealand Ltd, Wellington

First published 1979

© CSIRO 1979

Text set in 10/12 pt IBM Press Roman,
printed and bound at Griffin Press Ltd, Adelaide, SA

ISBN 0 273 08439 9

Contents

Preface

In 1975, when working as a Research Fellow at the Computer Centre of the Australian National University, I gave a course on Spectral Analysis as part of that year's teaching program for computer users. There were nine lectures given over three weeks, eight by myself and one by Dr D. Fraser of the CSIRO Division of Computing Research. The set of lecture notes handed out to those attending the course were later extensively revised and issued as Report R76-1 (1976) of the Department of Information Science, University of Tasmania, where I had taken up a lecturing position. In due course Dr Fraser arranged for the publication of the Report as a monograph by the Editorial and Publications Service of CSIRO. To this end a further revision was carried out, mainly to rectify a number of typographical errors and make minor improvements in the presentation. As this was nearing completion, the Service negotiated for the inclusion of the book in the present Series by Pitman, so that the book will be concurrently available in London. The editorial work and typesetting were handled by CSIRO in Melbourne.

The present book differs from other currently available texts on spectral analysis and digital signal processing in that it is neither orientated to engineering nor to statistics and no previous background in engineering mathematics, communication theory or probability is required. Our aim is to make spectral analysis techniques available to workers in any field. The book should be comprehensible to anyone with a basic working knowledge of calculus, although it would help it he also knows some statistics and matrix algebra. We apologize for the sketchiness of many mathematical derivations; the lecture notes were written with the idea that full derivations would be presented during the lectures themselves according to the needs of the students. In consequence, there may be a number of derivations in the book for which the reader will need to seek intermediary steps. However, let it be repeated, nothing more advanced than basic calculus will be called for.

It is a pleasure to acknowledge my gratitude to all those who contributed towards this work. First and foremost I wish to thank my friend and co-author Don Fraser for his time and effort, in giving his lecture, writing the notes and finally in getting the work published in book form. The book is as much his as mine. I also wish to thank Dr M. R. Osborne, Head of Computer Centre (now the Computing Research Group), Australian National University, for encouraging me to give the course, Dr R. S. Anderssen for discussions over its contents, and Professor A. H. J. Sale for his approval of the publication of the work, first as Report and then as book. Last but not least I must express my appreciation of the assistance of the CSIRO Editorial and Publications Service, particularly that of Mr B. J. Walby, Editor-in-Chief, without whose tireless effort the book would not have been available in its present form.

C. K. Yuen

July 1978

1 Introduction

The idea of 'spectrum' has been with us since the days of Newton, who decomposed sunlight into an array of coloured light using a prism. In due course, spectra of light and other types of electromagnetic radiation emitted by matter, both on earth and in the sky, became the most important indicators of their internal structures. Eventually, it was the inexplicable behaviour of some spectra which brought about the downfall of classical physics. Study of black body radiation ushered in Planck's energy quantum; hydrogen spectra led to Bohr's angular momentum quantization and finally 'modern' quantum mechanics. Michelson looked for ether drift with his interferometer, and found none; this was not explained satisfactorily until Einstein proposed relativity. Today's physicists still use the interferometer to measure spectra. (To be exact, the autocorrelation function, or Fourier transform of the spectrum.)

The measure and analysis of the spectra of other types of signals, however, did not become practicable until much later. They also took rather different forms. Light has very high frequencies. Our instruments cannot respond fast enough to directly measure the waveforms. Instead, we measure the amount of energy contained in the incoming signals. With the slower signals, such as mechanical vibrations, speech, cardiogram, population count, economic indices, etc., we can measure them as functions of time, and get the spectrum by computation. This is very nice in that more flexible modes of analysis are allowed. However, instruments sensitive enough to perform the measurements of some of these signals, and computing equipment fast enough to do the calculation, were not available in quantity until the 1940's for serious spectral analysis to start.

Briefly, the idea of spectral analysis is this: a given function $x(t)$ is expressed as a linear combination of a prescribed set of functions $y_i(t)$: $x(t) = \sum_i X_i y_i(t)$. Then, we define a set of quantities in terms of X which, in some sense or other, indicate the importance of $y_i(t)$ for representing $x(t)$. This, then, tells us something about the structure of $x(t)$, and where $x(t)$ is some signal produced by a physical, chemical, biological, engineering, economic or social system, we gain some understanding of the nature of the system. Thus, the spectrum of a cardiograph may tell the doctor whether the man producing it is healthy; the radiation spectrum shows the structure of atoms and molecules producing it; while the engineer might learn from the spectra of a signal function before and after its passage through some transmission medium how the distortions generated by the medium are to be combatted. And everybody would like to use the spectrum of the stock market index to predict future trends. (Few succeed, however.) For such problems, spectral analysis is only a very crude tool, despite the apparent complexity of some of the mathematical formulae and the large variety of techniques, and its theoretical foundation is still to some extent shaky. Hence, no one needs to feel surprise when the method fails. The surprise is, in fact, that it has worked so well so often.

In order that the spectrum serves the ultimate purpose of helping us understand the system producing the signal, the functions $y_i(t)$ must bear some relation to the structures inherently present in $x(t)$. This, alas, is not easy. The alternative is to use a set of functions that facilitate computation, hoping that they would still tell us something about the structures. This is why we almost always use the *sinusoidal* functions. These functions arise naturally from certain kinds of idealized systems called linear time-invariant systems, and give good measures of a type of 'stable regularity'. They are also quite convenient for calculation purposes. Consequently, they are fairly good in terms of cost and performance. Other function sets are rarely used.

It is by now clear that to be complete, spectral analysis has to have three parts. First, we must *define* the spectrum. That is, we must decide which set of functions $y_i(t)$ to use and why. Even when we just take the easy way out by using the sinusoids, it is necessary to know the reasons for making the choice. In addition, one must ask how the spectrum is to be defined in terms of X. The usual thing is to take the *mean square* of X_i as the measure of importance of $y_i(t)$. Only after the definition has been chosen can we actually devise procedures for *computing* the spectrum. And, after we obtain the results we have the problem of understanding what the spectrum tells us about the structure of the signal, and hence, that of the system producing it. So the last part of spectral analysis is that of *interpreting* the spectrum. Unfortunately, the first and third parts are very dependent on the particular applications. We cannot study them in any detail here. It is hoped, however, that our discussion of the computation part is sufficiently general that the reader will be able to adapt his knowledge to different needs. In any case, what we describe is the common practice, and should suit numerous applications.

The techniques for spectrum computation evolved gradually over the past thirty years, passing through some four stages of development. In the early days, people simply computed the Fourier transform of a signal function and tried to interpret it, looking for the important components. The quantity usually studied was actually the square of the transform, and at first it was plotted against period rather than frequency. As result, it was given the name 'periodogram'. The method was quickly discredited. It was soon realized that the periodogram behaves very erratically. Two portions of similar-looking signals produced by the same source under similar conditions can have completely different periodograms. Thus, most of the structures one sees in a periodogram are pure garbage. A component that looks prominent in one periodogram may be almost negligible in another, supposedly equivalent, periodogram. Consequently, one is quite unable to assign with confidence any physical meaning to the structures.

Another reason for the decline of the periodogram method was the time taken to compute the Fourier transform of a long piece of signal. It normally takes $\sim N^2$ multiplications and additions to Fourier transform N numbers, far too many for any sizable N in the days of Univac 1 and IBM 704.

The two problems were solved at one stroke by the correlation method for spectrum computation. It was noted that the periodogram can also be computed by Fourier transforming the autocorrelation of the input data. The autocorrelation was found to be fairly reliable for small values of argument but erratic for large arguments. It was naturally concluded that, instead of Fourier transforming the whole autocorrelation, we should take only its values for small arguments and cut off the bad parts. This not only gives more stable results; it is also much faster, since only a small part of the autocorrelation needs to be evaluated, say M values, and we require only an M-point Fourier transformation, taking only $\sim M^2$ operations.

It was also noticed that the computed spectrum could be further improved by averaging over neighbouring values in it, or alternatively, by multiplying the autocorrelation function into a function that reduces the input gradually and finally terminates it, rather than by abruptly cutting off the bad parts as previously done. The two improving operations were easily shown to be equivalent, and became known as windowing. In additon, windowing was shown to correspond to averaging over sets of values of the periodogram, a procedure which causes the erratic fluctuations to cancel out among themselves. In theory, therefore, one could have achieved the same result by computing the periodogram and carrying out smoothing operations directly. This would indeed produce good results. However, at the time it was thought this would be too time consuming.

Stage 3 of the development of spectral analysis was the idea of complex demodulation, which is basically the implementation of narrow band filters by computational means. The idea, however, hardly had time to take off before the fourth stage arrived. This was the re-discovery of the fast Fourier transform in 1965. By the use of FFT, it now takes only $\sim N(\log N)$ operations to Fourier transform N numbers. Therefore, all at once it became faster to Fourier transform a set of data, square to get its periodogram, and then smooth out the fluctuations to obtain a good spectrum, than to compute part of the autocorrelation. Thus, it is now fashionable to use the direct method rather than the indirect (autocorrelation) method, though in our opinion the latter still has a place, especially in preliminary experimentation.

The basic idea of spectrum computation, however, has not changed. Regardless of which technique we choose to employ, we must be sure that there are sufficient 'degrees of freedom' in the final computed results to ensure that they do not fluctuate too much. To be able to analyse such unpredictable fluctuations, we require some elementary knowledge of statistical modelling. Therefore, this book will start with an introduction to Fourier series, their computation, and the idea of random variables and random processes. After these topics we shall introduce the idea of power spectrum and its relation to the autocorrelation function, and discuss the problems we face in computing them from measured values of signals. The technique of windowing is then discussed. The next chapter summarizes spectrum computation by combining the discussion of previous chapters. This is followed by a chapter showing a few spectra computed from artificially generated data as well as real world signals. The last three chapters discuss three related topics.

As far as practical applications are concerned, the mathematics of spectral analysis is not difficult. However, the subject can be quite confusing because of the great variety of different techniques. We shall try to keep everything elementary and easy to understand, without sacrificing rigour completely.

In this short introduction we shall concentrate on what one calls 'univariate' spectral analysis, i.e., the study of the structure of a single piece of signal. Multivariate spectral analysis, which studies in addition the inter-relation between different pieces of signals, is at least equally important. This requires the computation of *cross-spectra*, on which we shall make a few brief comments. It is worth pointing out, however, that computing cross spectra is quite simple. It is their interpretation that is more difficult. To readers interested in pursuing the subject further we recommend the book of Jenkins and Watts.

2 Fourier Transform

Fourier series

The usual way to introduce the Fourier transform is to look at a periodic function, say with period T, defined over the real line $t \in [-\infty, \infty]$, and then extend the result to non-periodic functions by taking the limit $T \to \infty$. We shall not take this approach. In real life we can only look at signals of finite duration. It is also usual to present analysis of the convergence of a Fourier series, i.e., how closely it approaches the function it seeks to approximate as the number of terms tend to ∞, which is again impossible in practice. Instead of these, we shall merely look at the approximation of functions over a finite interval by a finite series. Most of the time we shall use the standard interval $[0, 1]$, since by appropriate shifts and scaling any finite interval can be converted into this one, but occasionally we shall use $[0, T]$ for the purpose of comparing intervals of different lengths.

Given $x(t)$ and the prescribed function set $y_i(t)$, we wish to find a set of coefficients X_i such that $\sum_i X_i y_i(t)$ is as close to $x(t)$ as possible. Now how do we define 'close'? The easiest thing to do is to require that the following quantity, called the *mean square error*, be minimized:

$$E = \int_0^1 |x(t) - \sum_i X_i y_i(t)|^2 \, dt. \tag{1}$$

(Note this is a time average, not an ensemble average.) The absolute value sign is needed here because X_i and $y_i(t)$ may be *complex*, even though $x(t)$ is real. We remind the reader that

$$|a|^2 = a^*a,$$

where a^* is the complex conjugate of a.

To find the X_i's, we differentiate E with respect to each X_i and set the derivative to 0:

$$\frac{\partial E}{\partial X_i} = 0 = 2 \int_0^1 \left(x(t) - \sum_j X_j y_j(t) \right) y_i^*(t) \, dt \, ,$$

where the index behind the summation sign was changed to j to distinguish it from i, i.e., we are summing over a set of indices j, but are differentiating with respect to one of them, i. If we assume that there are N values of X_i, then the above expression gives N equations in N unknowns:

$$\sum_j A_{ij} X_j = \int_0^1 y_i^*(t) x(t) \, dt \, , \tag{2}$$

with

$$A_{ij} = \int_0^1 y_i^*(t) y_j(t) \, dt \, . \tag{3}$$

(2) becomes much simpler if the y's are *orthonormal*, as defined by the following properties:

$$A_{ii} = \int_0^1 y_i^*(t) y_i(t) \, dt = 1 \, ,$$

$$A_{ij} = \int_0^1 y_i^*(t) y_j(t) \, dt = 0 \, , \qquad i \neq j \, . \tag{4}$$

This can be written as $A_{ij} = \delta_{ij}$. The symbol δ_{ij}, called the Kronecker delta, is just a concise notation for something which is unity (1) if $i = j$, and zero (0) if $i \neq j$. By substituting δ for A in eqn (2), $\sum \delta_{ij} X_j$ is just X_i, as $\delta_{ij} = 0$ except when $j = i$. Thus

$$X_i = \int_0^1 y_i^*(t) x(t) \, dt \, . \tag{5}$$

We then have a simple expression for finding X_i from $x(t)$.

Now let us choose as $y_i(t)$ the complex exponential functions:

$$y_i(t) = \exp(cit) \, , \tag{6}$$

where $c = 2\pi\sqrt{-1}$. (Note: we do not write $\sqrt{-1}$ as i, since we need all the letters i to n to stand for integers.) It is well known (see Appendix 1) that

$$\exp(2\pi\sqrt{-1}it) = \cos(2\pi it) + \sqrt{-1} \sin(2\pi it) \, , \tag{7}$$

and

$$\int_0^1 \exp(cit) \, dt = 0 \, , \qquad i \neq 0 \, , \tag{8}$$

which is easily proved by actually carrying out the integration. We can also see (8) by noting that, in the interval $[0, 1]$ there are i complete periods of $\sin(2\pi it)$ and $\cos(2\pi it)$, but as each is positive (+) over half of a period and negative (−) over the other half, the net integral is 0.

It then follows that

$$\int_0^1 y_i^*(t)\, y_j(t)\, dt = \int_0^1 \exp[c(j-i)t]\, dt = 0\,, \qquad j \neq i\,. \tag{9}$$

However, if $i = j$ then $\exp[c(j-i)t] = 1$, so

$$\int y_i^*(t)\, y_i(t)\, dt = 1\,. \tag{10}$$

Together, equations (9) and (10) give just (4). Thus, if we express our function as

$$x(t) \sim \Sigma\, X_i \exp(cit)\,, \tag{11}$$

then we would find X_i by

$$X_i = \int_0^1 \exp(-cit)\, x(t)\, dt\,. \tag{12}$$

Equation (11) is called the *Fourier series* for $x(t)$, and X_i is the *i*th *Fourier coefficient* of $x(t)$. The set of coefficients X_i, $i = 0, \pm 1, \pm 2, \ldots$, are together called the *finite Fourier transform* of $x(t)$. To compute X_i from $x(t)$ is to *Fourier transform* $x(t)$, or to perform a *Fourier transformation* on it. Index i is the *frequency* of X_i. Changing $x(t)$ by manipulating its Fourier coefficients is to operate in the *frequency domain*, or *Fourier transform space*, while changing $x(t)$ directly, perhaps in order to change its Fourier coefficients, is to operate in *time domain*. For the moment this is all the jargon we need.

Some properties of Fourier series

We may express as Fourier series both real and complex functions, but only real functions are of interest to us. We note, however, that even when $x(t)$ is real, X may be complex, as it has both a real part and an imaginary part. We know

$$\mathrm{Re}(X_i) = \int \mathrm{Re}[\exp(-cit)]\, x(t)\, dt = \int \cos(2\pi it)\, x(t)\, dt\,, \tag{13}$$

and

$$\mathrm{Im}(X_i) = \int \mathrm{Im}[\exp(-cit)]\, x(t)\, dt = \int \sin(2\pi it)\, x(t)\, dt\,. \tag{14}$$

If $x(t)$ is *even*, i.e., $x(1-t) = x(t)$, then $\mathrm{Im}(X_i) = 0$ as it is the integral of an even function multiplied into an odd function, $\sin(2\pi it)$. In this case X_i has no imaginary part. Thus, a real, even $x(t)$ has a real Fourier transform. On the other hand, if $x(t)$ is odd then the real part of X_i vanishes.

It is also clear that

$$X_i^* = \int_0^1 [\exp(-cit)]^* x^*(t)\, dt = \int \exp[-c(-i)t]\, x(t)\, dt = X_{-i}\,, \tag{15}$$

This holds whenever $x(t)$ is real (so that $x^*(t) = x(t)$). However, when $x(t)$ is also even then X_i is real, in which case we have

$$X_i = X_i^* = X_{-i}\,.$$

Thus, a real even function has a real, even Fourier transform. If $x(t)$ is odd then X_i is purely imaginary, so that

$$X_i = -X_i^* = -X_{-i} .$$

Or, a real, odd function has an imaginary, odd Fourier transform.

Let us take two functions $u(t)$ and $v(t)$, which are supposed to have error-free Fourier series:

$$u(t) = \sum_j U_j \exp(cjt) , \qquad v(t) = \sum_k V_k \exp(ckt) .$$

What is the Fourier series of their product $p(t) = u(t) v(t)$? Clearly,

$$\begin{aligned} P_i &= \int p(t) \exp(-cit) \, dt \\ &= \int \exp(-cit)[\sum_j U_j \exp(cjt)] \, [\sum_k V_k \exp(ckt)] \, dt \\ &= \sum_{j, k} U_j V_k \int \exp[-c(i-j)t] \exp(ckt) \, dt . \end{aligned}$$

The integral is 0 unless $i - j = k$, which gives

$$P_i = \sum_j U_j V_{i-j} . \tag{16}$$

This is called a *discrete convolution*. To take a convolution is to place two sets of numbers side by side, multiply each number from one set to its opposite number in the other set and sum the products, and then shift one set by a slot and repeat. Thus, P_0 is

$$\dots U_{-1} V_1 + U_0 V_0 + U_1 V_{-1} \dots ,$$

while P_1 is

$$\dots U_{-1} V_2 + U_0 V_1 + U_1 V_0 + U_2 V_{-1} \dots ,$$

etc. We thus have the statement 'Multiplication in the time domain (i.e., taking $u(t) v(t)$) corresponds to convolution in the frequency domain'.

A special case of interest is P_0. We have

$$P_0 = \int u(t) v(t) \, dt = \sum_j U_j V_{-j} = \sum_j V_j^* U_j ,$$

in view of (15). In particular, choosing $v = u$ gives

$$\int [u(t)]^2 \, dt = \sum_j | U_j |^2 . \tag{17}$$

This is known as Parseval's theorem.

An analogous problem is this: find

$$p(t) = \sum_i U_i V_i \exp(cit) ,$$

in terms of $v(t)$ and $u(t)$. Obviously,

$$p(t) = \sum_i U_i \exp(cit) \sum_j \delta_{ij} V_j$$

$$= \sum_i U_i \exp(cit) \sum_j V_j \int \exp(-cis) \exp(cjs) \, ds$$

$$= \int \sum_i U_i \exp[ci(t-s)] \sum_j V_j \exp(cjs) \, ds$$

$$= \int_0^1 u(t-s) \, v(s) \, ds \, . \tag{18}$$

Equation (18) is analogous to (16). It is a *continuous convolution*. Evaluation of $p(t)$ involves shifting one function by t, multiplying point by point into the other function, and then integrating the product. Equation (18) is described by the statement 'multiplication in frequency domain corresponds to convolution in time domain'. But there is a problem. In (18), as s goes from 0 to 1, $(t-s)$ can become negative. How do we define $u(t-s)$ then? To answer this question, let us trace back to the step before (18), where we put

$$\sum_i U_i \exp[ci(t-s)] = u(t-s) \, . \tag{19}$$

We note that $\exp(cit)$ is periodic if we extend t outside $[0,1]$, since

$$\exp[ci(t+1)] = \exp(cit) \exp(2\pi\sqrt{-1}) = \exp(cit) \, ,$$

as (Appendix 1)

$$\exp(2\pi\sqrt{-1}) = \cos(2\pi) + \sqrt{-1} \, \sin(2\pi) = 1 \, .$$

Thus, when $t-s < 0$ the left hand side of (19) is equal to

$$\sum_i U_i \exp[ci(t-s+1)] = u(t-s+1) \, .$$

In other words, when computing the convolution (18), if the argument goes outside the limits $[0,1]$, we simply assume that $u(t)$ is periodically extended. This makes (18) a cyclic convolution. (When you leave one end of the interval, you re-enter at the other end, as if you are on a circle.) In effect, (18) should be written as

$$p(t) = \int_0^t u(t-s) \, v(s) \, ds + \int_t^1 u(t-s+1) \, v(s) \, ds \, .$$

However, it would be good enough to keep to the form of (18) provided we keep in mind the periodic extension idea.

In the above expression, we could change the variable of integration to $s' = t-s$ in the first term, and to $s' = t-s+1$ in the second term. This turns it into

$$p(t) = \int_0^t u(s') \, v(t-s') \, ds' + \int_t^1 u(s') \, v(t-s'+1) \, ds' \, .$$

This time, it is $v(t)$ that has been cyclically shifted. In other words, in a cyclic convolution we could consider either function to have been shifted and to have been periodically extended.

Gibbs phenomenon

Let us consider the truncated Fourier series of function $x(t)$:

$$x^*(t) = \sum_{i=-M}^{M} X_i \exp(cit). \tag{20}$$

By taking the lower limit at the negative value of the upper limit, we assure ourselves that the result is real, since the imaginary part of $X_i \exp(cit)$ cancels that of $X_{-i} \exp(-cit)$. To find $x^*(t)$ in terms of $x(t)$, we note that (20) can be considered as a special case of (18), with $U_i = X_i$ and $V_i = 1$, $-M \leqslant i \leqslant M$, $= 0$ for other i. To find $v(t)$, we have

$$v(t) = \sum_{i=-M}^{M} \exp(cit) = \sum_{i=-M}^{M} [\exp(ct)]^i .$$

This is a geometric series of $2M+1$ terms, with leading term $[\exp(ct)]^M$ and each succeeding term being $\exp(-ct)$ times the previous term. Such a series is readily summed by the expression

$$S_n = a_0 + a_1 + ... + a_n = a_0(1 - f^{n+1})/(1 - f) ,$$

where f is the inter-term ratio, a_{i+1}/a_i, in this case $\exp(-ct)$, and n is the number of terms minus one, in this case $2M$. Thus we have

$$v(t) = \exp(cMt)\{1 - \exp[-c(2M+1)t]\}/[1 - \exp(-ct)]$$

$$= \{\exp[\tfrac{1}{2}c(2M+1)t] - \exp[-\tfrac{1}{2}c(2M+1)t]\}/[\exp(\tfrac{1}{2}ct) - \exp(-\tfrac{1}{2}ct)]$$

$$= \sin[(2M+1)\pi t]/\sin(\pi t) , \tag{21}$$

which arises after we make use of (7). This function has the form shown in Fig. 2.1. Note that we have shown its periodic extension outside the

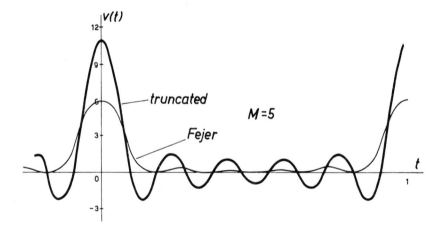

Fig. 2.1

interval $[0, 1]$. Clearly, $v(t) = 0$ at $t = k/(2M+1)$, $k = 1, 2, ..., 2M$, but its value at 0 and 1 is $2M+1$, which is found by applying the l'Hospital rule. (That is, near $t = 0$ $\sin(\pi t) \sim \pi t$ and $\sin[(2M+1)\pi t] \sim (2M+1)\pi t$.) As t moves towards $t = \frac{1}{2}$ we get a series of peaks of alternating signs and decreasing height. The peak centred on $t = 0$ has base width $2/(2M+1)$, while the others have half that width. Further,

$$\int_0^1 v(t)\,dt = \Sigma \int \exp(cit)\,dt = 1\,, \tag{22}$$

since only the $i = 0$ term gives a non-zero contribution.

The $v(t)$ given by (21) appears frequently in our discussion, and we shall give it the special notation $D_M(t)$. Using (18) we have

$$x^*(t) = \int_0^1 D_M(s)\, x(t-s)\,ds\,. \tag{23}$$

(Remember the periodic extension!) Thus, $x^*(t)$ is a sort of weighted integral of $x(t)$. If $x(t)$ is continuous and varies slowly, then most of the alternating peaks tend to cancel, so that $x^*(t)$ is approximately just $x(t)$ averaged over an area covered by the main peak. Since $x(t)$ varies slowly, its average over a short interval does not differ appreciably from the exact values, so that $x^*(t)$ would approximate $x(t)$ well. However, if $x(t)$ contains a discontinuity, then its integral with $D_M(t)$ would oscillate rapidly in the neighbourhood of the discontinuity. Fig. 2.2 shows such a function and its

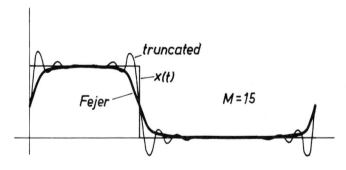

Fig. 2.2

Fourier series approximation. Not only is the error large near the discontinuity, we have also the fact that error does not decrease if we take a larger M. The peaks in the error curve get narrower as M increases, but remain at about the same values. This is why mathematicians say 'Fourier series do not converge uniformly, only in the mean square sense', which means that, the error does not go to zero everywhere as the series gets larger, but if you square the error and integrate over the whole interval, the result does go to zero. (This is because the error curve peaks get narrower all the time.) The failure of Fourier series to converge at discontinuities is given the name 'Gibbs phenomenon'.

In practical computation we can only handle finite series. Consequently, we do not 'see' the exact function $x(t)$ if we obtain it from its Fourier transform. Instead, we 'see' the approximation $x^*(t)$, which differs from $x(t)$ because of its finite extent in the frequency domain. We express this by saying that $x(t)$ is 'seen' through a *window*. The window $D_M(t)$ operates on $x(t)$ in the time domain, and so is called a *time window*. It corresponds to a *frequency window*, which is 1 for $-M \leqslant i \leqslant M$ and 0 elsewhere. This is called a rectangular window or boxcar window. There is a one-to-one correspondence between time windows and frequency windows. Multiplication by a window in one domain gives rise to convolution in the other domain.

The convergence of a finite Fourier series is considerably improved if we replace the rectangular window by a triangular window:

$$V_i = 1 - |i/(M+1)|, \qquad -M \leqslant i \leqslant M,$$
$$= 0 \text{ for other } i. \tag{24}$$

In other words, we approximate $x(t)$ by a Fejer series,

$$x^\#(t) = \sum_{i=-M}^{M} X_i[1 - |i/(M+1)|] \exp(cit). \tag{25}$$

It is useful to note

$$1 - |i/(M+1)| = \sum_{j=|i|}^{M} \frac{1}{M+1}.$$

To find $v(t)$, we take

$$v(t) = \sum_{i=-M}^{M} \left(\sum_{j=|i|}^{M} \frac{1}{M+1} \right) \exp(cit) = \frac{1}{M+1} \sum_{j=0}^{M} \sum_{i=-j}^{j} \exp(cit).$$

On comparing the second sum with (21) we see that

$$v(t) = \frac{1}{M+1} \sum_{j=0}^{M} \sin[(2j+1)\pi t]/\sin(\pi t)$$

$$= \frac{1}{(M+1)\sin(\pi t)} \sum_{j=0}^{M} \{\exp[c(2j+1)t] - \exp[-c(2j+1)t]\}/2\sqrt{-1}.$$

(Remember $\exp(c\sqrt{-1}\,t) = \cos t(2\pi t) + \sqrt{-1}\,\sin(2\pi t)$.) The two terms in the sum form two geometric series, and so can be easily computed as before. After some long manipulation similar to that preceding (21) we get

$$v(t) = \sin^2[(M+1)\pi t]/\sin^2(\pi t)(M+1). \tag{26}$$

This is plotted in Fig. 2.1. It is clearly much less oscillatory than $D_M(t)$. Fig. 2.2 shows the same function seen before approximated by a Fejer series. The convergence is clearly more even. It should be said, however, that the mean square error is larger, since we are no longer using the mean square error minimizing coefficients.

Leakage

We saw that, an abrupt termination of the Fourier series in frequency domain, as in (20), causes serious convergence problems in time domain. A similar phenomenon exists in reverse form: an abrupt cutoff of a function in time domain gives an oscillatory Fourier transform. This is not, however, called the Gibbs phenomenon, but rather, 'leakage', for the reasons we shall see below.

Let us consider the function $x(t) = 0$, $0 \leqslant t \leqslant \Delta$, and $1-\Delta \leqslant t \leqslant \Delta$, $= \exp(cjt)$, $\Delta < t < 1-\Delta$, i.e., a complex exponential terminated abruptly at both ends. Its Fourier coefficients are:

$$X_i = \int_{\Delta}^{1-\Delta} \exp(cjt) \exp(-cit) \, dt$$

$$= \{ \exp[c(j-i)(1-\Delta)] - \exp[c(j-i)\Delta] \}/[c(j-i)]$$

$$= \exp[\tfrac{1}{2}c(j-i)] \{ \exp[c(j-i)(\tfrac{1}{2}-\Delta)] - \exp[c(j-i)(\Delta-\tfrac{1}{2})] \}/c(j-i)$$

$$= \exp[\tfrac{1}{2}c(j-i)] \, \sin[\pi(1-2\Delta)(j-i)]/[\pi(j-i)] \, , \qquad (27)$$

compared with

$$\int_0^1 \exp(cjt) \exp(-cit) \, dt = \delta_{ij} \, .$$

That is, while the Fourier transform of $\exp(cjt)$ should have only one non-zero term, at $i = j$, if we terminate it abruptly we would see additional non-zero terms. Thus, the Fourier transform has 'leaked' from the $i = j$ term to elsewhere. Comparison of (27) with (21) shows that the leaked terms are distributed in a way rather similar to Gibbs phenomenon error peaks.

Even when there is no abrupt termination of $x(t)$ leakage can still arise. Let us now consider $x(t) = \exp[c(j+a)(t+b)]$, an exponential with frequency $j+a$ and *phase b*, $0 \leqslant a, b \leqslant 1$. The phase b merely causes a shift along the time axis of the whole function. We find that

$$X_i = \exp[c(j+a)b] \int_0^1 \exp[c(j+a-i)t] \, dt \, .$$

After similar manipulations to those above we get

$$X_i = \exp[c(j+a)b + \tfrac{1}{2}c(j+a-i)] \, \sin[\pi(i-j+a)]/[\pi(i-j+a)]$$

Since b is arbitrary, let us choose $b = -\tfrac{1}{2}$. This gives

$$X_i = \exp(-\tfrac{1}{2}ci) \, \text{sinc}\,(i-j+a) \, , \qquad (28)$$

where we have defined a function similar to $D_M(t)$, both in shape and in importance,

$$\text{sinc}\,(t) = \sin(\pi t)/(\pi t) \, .$$

First, let us take $a = 0$. Obviously, $\text{sinc}\,(i-j) = 0$ if $i \neq j$, since $\sin[\pi(i-j)] = 0$. For $i \neq j$, use of l'Hospital rule gives $\text{sinc}(0) = 1$. That is, if $a = 0$ we have

$$X_i = \exp(-\tfrac{1}{2}ci)\delta_{ij} \,.$$

This just gives back the known result, namely, that a complex exponential with integer frequency has only one non-zero Fourier coefficient. (The factor in front comes from the non-zero phase, $-b = \tfrac{1}{2}$.) But if $a \neq 0$ then $\text{sinc}\,(i-j+a) \neq \delta_{ij}$, and every X_i is non-zero. We see that, an exponential with frequency somewhere between two integers (j and $j+1$) has a Fourier transform which is not confined to these two integers. It, too, 'leaks out'.

The problem of leakage is potentially a very serious one. When we compute spectra, we wish usually to identify peaks in them. A peak in the Fourier transform corresponds to an exponential with some frequency. However, because of leakage we do not see just a peak. Rather, we see a series of peaks of varying sizes. This is bad not only because the peak we look for is changed, but more seriously because the leakage to other frequencies can completely obscure the structures actually present there. Similarly, a sharp change in the Fourier transform will generate through leakage a series of oscillations similar to Gibbs phenomenon errors. (If, however, the Fourier transform is slowly varying, then leakage is not a problem at all. We can consider such a transform as a series of continuous peaks of comparable height, all leaking out. But because leaking is highly oscillatory, the contribution from different peaks tend to cancel out among themselves, leaving little net leakage.)

Can we prevent leakage? The answer is no. We can only replace bad leakage by less bad leakage. Instead of allowing the signal to leak all over the place, as in (27) and (28), we try to confine leakage to frequencies close to the true frequency, while sharply reducing leakage to frequencies farther away. The techniques for doing this are generally called 'windowing'. As in the previous section, leakage makes it impossible for us to 'see' the true Fourier transform. Instead, we see a different one obscured by a window. The solution is then to design a better window, an art generally known as 'window carpentry'. However, we shall leave this subject until the reader has been introduced to a few more of the elementary concepts.

The discrete Fourier transform

We have previously defined the finite Fourier transform of a function as an integral

$$X_i = \int_0^1 x(t) \exp(-cit)\,\mathrm{d}t \,.$$

In actual computation we can only use a finite number of measured values of $x(t)$. Let us suppose we measure, or *sample* as is often said, $x(t)$ at

$t = j\Delta$, $j = 0, 1, ...$, where Δ is the time interval between two *sampled values* of x and is called the *sampling interval*. If we require N sampled values, it is natural to take $\Delta = 1/N$. Thus we approximate X_i by the sum

$$\hat{X}_i = N^{-1} \sum_{j=0}^{N-1} x(j/N) \exp(-cij/N) . \tag{29}$$

This is called the N-point discrete Fourier transform (DFT). Note that dt has been replaced by Δ, or N^{-1}.

How good is the approximation? It is usually good, sometimes very good. To explain this, let us consider an $x(t)$ that is known to have non-zero Fourier coefficients X_k only for $-\tfrac{1}{2}N < k < \tfrac{1}{2}N$. (If N is odd, k lies between $-\tfrac{1}{2}N+\tfrac{1}{2}$ and $\tfrac{1}{2}N-\tfrac{1}{2}$; if N is even, between $-\tfrac{1}{2}N+1$ and $\tfrac{1}{2}N-1$, inclusive.) Thus, $x(t)$ can be expressed as

$$x(t) = \sum_{-\frac{1}{2}N < k < \frac{1}{2}N} X_k \exp(ckt) . \tag{30}$$

Now we compute its N-point DFT for $-\tfrac{1}{2}N < i < \tfrac{1}{2}N$:

$$\hat{X}_i = N^{-1} \sum_{j=0}^{N-1} \exp(-cij/N) \sum_k X_k \exp(ckj/N)$$

$$= N^{-1} \sum_k X_k \sum_{j=0}^{N-1} \exp[c(k-i)j/N] . \tag{31}$$

The summation over j is again a geometric series, with leading term $\exp(0) = 1$ and inter-term ratio $\exp[c(k-i)/N]$. The total is

$$\frac{1 - \{\exp[c(k-i)/N]\}^N}{1 - \exp[c(k-i)/N]} = \frac{\exp[\tfrac{1}{2}c(k-i)(N-1)/N] \sin[\pi(k-i)]}{\sin[\pi(k-i)/N]} \tag{32}$$

If $k = i$, the expression above is N. (L'Hospital's rule again.) If $k \neq i$ the total is 0, as long as $(k-i)/N$ is not an integer, which is satisfied here because $-\tfrac{1}{2}N < i, k < \tfrac{1}{2}N$ so that their difference cannot be as large as N. In other words, (32) in this case is the same as $N\delta_{ki}$. Hence (31) is

$$\sum_k X_k \delta_{ki} = X_i .$$

In short, for this particular $x(t)$ the discrete Fourier transform is exactly the same as the finite Fourier transform. Since we are able to compute all the non-zero X's exactly from N sampled values, we can just substitute them into (30), which gives $x(t)$ for any t. That is to say: if a function has a finite Fourier transform confined to within the indices $-\tfrac{1}{2}N$ and $\tfrac{1}{2}N$ exclusive, then it can be exactly reconstructed from N sampled values taken at $t = j/N$, $j = 0, 1, ..., N-1$. This is known as the sampling theorem.*

---

* The usual discussion assumes $x(t)$ to be defined over the complete real line. An important point of difference is discussed in the next section.

We note that the sampling interval, Δ, is just $1/[2(\tfrac{1}{2}N)]$ and $\tfrac{1}{2}N$ is the frequency limit of x. In engineers' language it is said that 'if a function contains harmonic components of frequency less than f only, then sampling at intervals of $1/2f$ is sufficient for exactly reconstructing the function'.

Now let us look at a general function $x(t)$ and see how it is related to its DFT series $\hat{x}(t)$, defined as $\Sigma\, \hat{X}_i \exp(cit)$. More exactly

$$\hat{x}(t) = \sum_{i=-M}^{M} \exp(cit) . N^{-1} \sum_{j=0}^{N-1} \exp(-cij/N)\, x(j/N) ,$$

M to be specified later. By summing over i first and applying (21) we get

$$\hat{x}(t) = N^{-1} \sum_j x(j/N)\, D_M(t-j/N) . \tag{33}$$

Thus $\hat{x}(t)$ is an approximation to $x(t)$ *constructed purely from sample values of* $x(t)$. In other words, we are approximating $x(t)$ by a type of interpolation. We know that, if $x(t)$ has a restricted Fourier series as in (30), then $\hat{X}_i = X_i$. We can then make $\hat{x}(t) = x(t)$ for all t. Even when $x(t)$ does not satisfy the restriction (30), we still have $\hat{x}(t) = x(t)$ for a discrete set of values of t. First we assume N to be odd. By choosing $2M+1 = N$, we have by definition of D_M, (21),

$$N^{-1} D_M(t-j/N) = 1 , \quad t = j/N; \quad = 0 , \quad t = k/N , \qquad k \neq j .$$

This ensures $\hat{x}(k/N) = \Sigma_j\, x(j/N)\, \delta_{kj} = x(k/N)$, or that $\hat{x} = x$ at the sample points. However, it may not be exact at other values of t.

Unfortunately, we can choose $2M+1$ to be N only when N is odd. For even N things are a little more complicated. We would like M to be $\tfrac{1}{2}N-\tfrac{1}{2}$, but since this is not permitted we are forced to use two values, $\tfrac{1}{2}N$ and $\tfrac{1}{2}N-1$, and average the two series. This gives

$$\hat{x}(t) = N^{-1} \sum_j x(j/N)[D_{\tfrac{1}{2}N}(t-j/N) + D_{\tfrac{1}{2}N-1}(t-j/N)]$$

$$= N^{-1} \sum_j x(j/N) \,\tfrac{1}{2}\{\sin[(N+1)\,\pi(t-j/N)] + \sin[(N-1)\,\pi(t-j/N]\}$$

$$/\sin[\pi(t-j/N)] .$$

Using the trigonometric identity $\sin(a \pm b) \equiv \sin a \cos b \pm \cos a \sin b$ we get

$$\hat{x}(t) = N^{-1} \sum_j x(j/N) \sin[N\pi(t-j/N)]/\tan[\pi(t-j/N)] .$$

(Note we used the relation $\tan = \sin/\cos$.) It is easily verified that we still have

$$\hat{x}(k/N) = \sum_j x(j/N)\, \delta_{kj} = x(k/N) .$$

The foregoing longwinded argument shows that given the N-point DFT of a function, we can always construct a Fourier series that recovers the function exactly at the sample points; if the function has a restricted Fourier transform, then we can reconstruct it exactly everywhere. Hence the earlier statement: usually good; sometimes very good.

Fig. 2.3 illustrates the so-called process of *sinusoidal interpolation*, the reconstruction of $x(t)$ from its sampled values as a sum of values of D. It is, of course, possible for many functions to be equal at the sampling points but different elsewhere. They would give the same sinusoidal interpolation. The interpolation would reproduce exactly the function that has a restricted Fourier series, but would differ from the other functions. It is fairly easy to see how significant this difference is, as we show next.

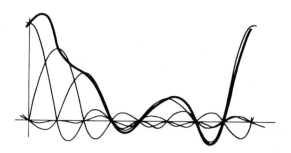

Fig. 2.3

Aliasing

Next let us look at $\hat{X}$ for a general $x(t)$. We already proved in (32) that

$$\hat{X}_i = \sum_k X_k \delta'_{ik} \, ,$$

where $\delta'_{ik} = 0$ unless $(k-i)/N$ is an integer, in which case it is 1. As we saw, if i and k are restricted to the values $-\tfrac{1}{2}N < i, k < \tfrac{1}{2}N$, there is no difference between δ and δ' as $i-k < N$. Now that the restriction has been removed, we have

$$\delta'_{ik} = 1 \, , \qquad k = i+jN \, , \qquad \text{for any integer } j; \quad \delta'_{ik} = 0 \text{ otherwise.}$$

This leads to

$$\hat{X}_i = \sum_{j=-\infty}^{\infty} X_{i+jN} \, . \tag{34}$$

Let us look at two values of x, $\exp(ckt)$ and $\exp[c(k+N)t]$. For the first, $X_k = 1$ (all other $X = 0$); in the second $X_{k+N} = 1$ (all other $X = 0$). When we compute DFT, we find that in both cases

$$\hat{X}_k = \ldots + X_{k-N} + X_k + X_{k+N} + \ldots = 1 \, .$$

In fact, the DFT of the two functions are exactly the same. The reason is quite obvious: at the sample points $t = j\Delta = j/N$ we have

$$\exp[c(k+N)j/N] = \exp(ckj/N)\exp(ckj) \, ,$$

and the second term is always 1. Thus, $\exp(ckt)$ and $\exp[c(k+N)t]$ are indistinguishable when we sample at intervals of $\Delta = 1/N$. Consequently, all those X with indices $k+jN$ get lumped together into the term $\hat{X}_k$. This indistinguishability of different functions under sampling is known as *aliasing*. We also talk of the frequency range $[-\infty, \infty]$ being *folded* into a low frequency range, i.e., $[N, 2N]$, $[2N, 3N]$, etc., are all added to $[0, N]$.

We have looked at complex exponentials. Things are a little more complicated for real functions. Let $x(t)$ be $\cos(2\pi jt)$, $\frac{1}{2}N < j < N$. This is of course identical to $\frac{1}{2}\exp(cjt) + \frac{1}{2}\exp(-cjt)$, which is indistinguishable from $\frac{1}{2}\exp[c(j-N)t] + \frac{1}{2}\exp[-c(j-N)t] = \cos[2\pi(N-j)t]$. Similarly, $\sin(2\pi jt)$ is indistinguishable from $\sin(2\pi(N-j)t]$. In other words, a real function with some frequency in $[\frac{1}{2}N, N]$ appears under sampling to have frequency $N-j$, not $j-N$ as in the case of complex exponentials. Suppose we have an $x(t)$ that contains non-zero frequency components up to $j = N-m$, m being a small integer, then the frequencies $[\frac{1}{2}N, N-m]$ are folded into $[m, \frac{1}{2}N]$, while frequencies in $[0, m]$ are not affected by aliasing at all.

Fig. 2.4

Fig. 2.4 illustrates the aliasing effect for sinusoidal functions. With only 5 samples over $[0, 1]$, a function of frequency 4 appears identical to one with frequency 1. Fig. 2.5 shows how a TV test pattern alters its appearance when we enlarge the sampling interval first in the x direction, then in the y direction, and finally both. Note the reversal of some frequency intervals. The reason for this is shown below.

The above are only theoretical results. In practice one never gets a function with absolutely no high frequency components. Even when a signal has been filtered to remove high frequency parts, we still get some high frequency because of *leakage*, which converts a non-integer low frequency component into a sum of integer frequency (both high and low) components. Aliasing then converts the high frequency components back into low frequency ones. In this way, a non-integer low frequency component appears under sampling as a combination of integer low frequency components. The discrete Fourier transform gives us the exact Fourier coefficients only when $x(t)$ satisfies (30), i.e., it contains only *integer* low frequency components. The usual discussion of DFT tends to ignore the 'integer' property. This is because it is assumed that $x(t)$ is defined over the complete real line. Since we do not have infinitely long functions in

Fig. 2.5

practice, the sampling theorem as it is usually stated is of little real relevance. It is often said that 'we can avoid aliasing by filtering out the high frequency parts before sampling'. This statement is clearly invalid, since analog filtering would not remove the non-integer low frequency components. Whenever there is leakage, there is aliasing; and since it is impossible to eliminate leakage, it is impossible to eliminate aliasing.

However, as we saw on pages 13-14, leakage decreases as we move away from the true frequency. If $x(t)$ contains only true frequencies of less than f, then even with leakage it is still reasonable to say that it contains no integer frequency above $1.25f$, and that sampling at intervals of $1/(2.5f)$ will recover the true Fourier transform fairly accurately. This is the recommendation usually given in textbooks. However, if the true Fourier transform of $x(t)$ contains a sharp peak just below f, the leakage into high frequencies would be particularly large, in which case it may be necessary to sample at smaller intervals, e.g., $1/4f$.

Now let us consider a real signal with restricted frequency range $[nf, (n+1)f]$. If we sample at $\Delta = 1/2f$, then the range is folded into $[0, f]$. Since there are no other frequencies present, there is no aliasing error. Every frequency component retains its correct amplitude, except that it appears to be low frequency. As result, the discrete transform will recover all X values correctly (provided there is negligible leakage). There is, however, one important point to remember. As shown a little while ago, a sinusoid of frequency ϕ, $f < \phi < 2f$, appears as one of frequency $f-\phi$ under sampling. Thus, the range $[f, 2f]$ is folded into $[f, 0]$, not $[0, f]$. On the other hand, when $n = 0$ there is of course no aliasing at all, so that $[0, f]$ is just $[0, f]$. It is not difficult to see that, whenever n is even there is no frequency reversal, but there is reversal for any odd n.

To put it differently, we find that for a real function of maximum frequency f, sampling at intervals of $\Delta = 1/2f$ will recover *all* its Fourier components correctly; sampling at a larger Δ will give some of its co-efficients correctly. If we write $1/2\Delta$ as f', N is just $2f'$. The maximum frequency f is aliased into $2f'-f$. Provided Δ is not too small, the frequencies $[0, 2f'-f]$ are unaffected by aliasing.

One point might strike the reader as being curious: when we compute an N-point DFT, almost half of the results produced are redundant. That is, $\hat{X}_i$ for $i = \frac{1}{2}N+1, \frac{1}{2}N+2, ..., N-1$, are equal to $\hat{X}_{i-N}$, i.e., same as $\hat{X}_{i'}$ for $i' = -\frac{1}{2}N+1, -\frac{1}{2}N+2, ..., -1$. Now, $X_i^* = X_{-i}$. Thus, $\hat{X}_i = \hat{X}_{i-N} = \hat{X}_{N-i}^*$ for $\frac{1}{2}N+1 < i < N-1$. Why, then, do we not compute $\hat{X}_0$ to $\hat{X}_{\frac{1}{2}N}$ only, and obtain the others by complex conjugation? We shall see in the next chapter that this is not possible if we wish to use the fast Fourier transform. Given N values of x, we have to proceed as if we need all N values of X. There is, however, a trick that gets around the problem. This is also shown in chapter 3.

Fourier transformation over [0, T]

Over [0, 1] we have

$$X_i = \int_0^1 \exp(-cit)\, x(t)\, dt\,,$$

and

$$x(t) = \sum_i X_i \exp(cit)\,.$$

We now wish to re-define everything on [0, T] instead. This is done by replacing t by t'/T, giving

$$\bar{x}(t') = x(t'/T) = \sum_i \bar{X}_i \exp(cit'/T)\,,$$

and

$$\bar{X}_i = \int_0^T \exp(-cit'/T)\, x(t'/T)\, dt'/T = \frac{1}{T}\int_0^T \exp(-cit/T)\, \bar{x}(t)\, dt\,.$$

In other words, if we wish to express a function $\bar{x}$ defined over [0, T] as a Fourier series, we would compute the coefficients $\bar{X}_i$ by integrating $\bar{x}(t)$ with $\exp(-cit/T)$, not with $\exp(-cit)$. This is an exponential function with frequency i/T rather than i.

Let us suppose we wish to evaluate $\bar{X}$ using sampled values, for $t = 0$, T/N, $2T/N$, ..., or jT/N. Δ is of course T/N. This turns the above into

$$\bar{X}_i = \frac{\Delta}{T}\sum_j \exp[-ci(jT/N)/T]\,\bar{x}(jT/N)$$

$$= N^{-1}\sum_j \exp(-cij/N)\, x(jT/N)\,. \tag{35}$$

The exponential function values used in (35) are the same as those used in (29).

Here we have a very pleasant feature of DFT: when we obtain N sample values of x over [0, T] and then perform an N-point DFT, *the frequencies we use are automatically the correct ones.* We do not have to worry about choosing them, scaling by T, etc.

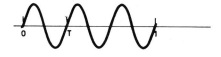

Fig. 2.6

The above arises not because of good luck. Regardless of what interval we use, the exponential functions are really the same set. The 0th function is a constant, the 1st a function that has one complete period over the interval, and the ith function has i complete periods. We have merely stretched or compressed the functions along the t axis. Fig. 2.6 illustrates this. A function that has one complete period of oscillation over $[0, T]$ for small T appears to have a larger frequency if we look at it over $[0, 1]$, and a function with one complete period over $[0, T]$ for large T appears to have a small frequency, consistent with the inverse relation i/T.

A different way of looking at the situation is as follows. When we have a large interval $[0, T]$, we are getting more data. Hence, we need to look at more frequencies by spacing them closer together. When we have a small interval, we can only look at fewer, more widely spaced frequencies.

It must be emphasized that, when we increase T we do not obtain a wider spectrum. Rather, we get a more densely spaced spectrum. As we saw on page 20, the maximum frequency we can look at is determined by the sampling interval and is completely independent of T. The fact is also clearly shown by our integrating $x(t)$ with $\exp[-cit/T]$ over $[0, T]$. With larger T, the difference between the frequencies of $\exp[-cit/T]$ and $\exp[-c(i+1)t/T]$ decreases. Thus, it would take more Fourier coefficients to cover the same frequency range. This is why we do not have to worry much about the interval over which the Fourier transform is defined. We can always assume that our interval is $[0, 1]$, with the recognition that the frequency i really corresponds to frequency i/T, where T is the length of the interval.

3 Fast Fourier Transform

Derivation

The fast Fourier transform (FFT) is an *algorithm* for computing the DFT of a function with minimum computational effort. The method became well known after the publication of the paper of Cooley and Tukey (*Mathematics of Computation*, 1965, **19**, 297–301), although it had been used in various forms by others before this. The algorithm is available as subroutines in many mathematical software packages, and it is not expected that the reader will need to program this for himself. However, we wish to give the reader some familiarity with the mathematical basis of the FFT and its peculiarities so that he will be able to make the best use of it.

We recall

$$\hat{X}_i = \sum_{j=0}^{N-1} \exp(-cij/N)\, x(j/N)\,, \tag{1}$$

is the N-point DFT of x. This is the multiplication of an $N \times N$ matrix, which is often written as W_N, into an N-element vector. The matrix is peculiar in that it contains only N different values, for $ij = 0, 1, ..., N-1$. When $ij = N$ $\exp(cij/N) = \exp(2\pi\sqrt{-1}) = 1$. In fact, whenever $ij > N$ we can apply the equality

$$\exp(-cij/N) = \exp[-c(ij-N)/N]\,, \tag{2}$$

i.e., subtract integer multiples of N from ij, until we obtain one of the N values already listed. For convenience we shall write $\exp(-ck/N)$ as $\bar{k}$ for $0 \leqslant k \leqslant N-1$. Thus, $\exp(-cij/N)$ for $i = 2$ and $j = 3$ is just $\bar{6}$, and for either $i = 0$ or $j = 0$ $\exp(-cij/N) = \bar{0}$. We know from (2) that $\exp(-ck/N)$ for $k \geqslant N$ is just $\overline{k-N}$. Note that $\bar{0} = 1$ and $\overline{\tfrac{1}{2}N} = -1$. It is also easy to show that

$$\overline{k+\tfrac{1}{2}N} = \bar{k} \exp(\pi\sqrt{-1}) = -\bar{k}\,, \tag{3}$$

and

$$\overline{N-k} = \exp[c(k-N)/N] = \bar{k}^* . \qquad (4)$$

With this notation, the matrix W_8 has the 64 elements listed below

$$
\begin{vmatrix}
\bar{0} & \bar{0} & \bar{0} & \bar{0} & \bar{0} & \bar{0} & \bar{0} & \bar{0} \\
\bar{0} & \bar{1} & \bar{2} & \bar{3} & \bar{4} & \bar{5} & \bar{6} & \bar{7} \\
\bar{0} & \bar{2} & \bar{4} & \bar{6} & \bar{0} & \bar{2} & \bar{4} & \bar{6} \\
\bar{0} & \bar{3} & \bar{6} & \bar{1} & \bar{4} & \bar{7} & \bar{2} & \bar{5} \\
\bar{0} & \bar{4} & \bar{0} & \bar{4} & \bar{0} & \bar{4} & \bar{0} & \bar{4} \\
\bar{0} & \bar{5} & \bar{2} & \bar{7} & \bar{4} & \bar{1} & \bar{6} & \bar{3} \\
\bar{0} & \bar{6} & \bar{4} & \bar{2} & \bar{0} & \bar{6} & \bar{4} & \bar{2} \\
\bar{0} & \bar{7} & \bar{6} & \bar{5} & \bar{4} & \bar{3} & \bar{2} & \bar{1}
\end{vmatrix}
\begin{matrix} \\ \leftarrow \\ \\ \\ \\ \leftarrow \\ \\ \\ \end{matrix}
\qquad (5)
$$

Let us compare rows i and $i+\tfrac{1}{2}N$, e.g., 1 and 5 as marked with arrows. We find that the elements for even j are identical; those for odd j differ by the same factor of 4. Further, if we confine our attention to the ith row, we see that the even j elements differ from neighbouring odd j elements by a constant factor of $\bar{i}$. Thus, there is a natural grouping of rows into $i < \tfrac{1}{2}N$ and $i \geqslant \tfrac{1}{2}N$, and columns into even and odd j's. Another illustration of such relations is given in Fig. 3.1, where we plot $\cos(2\pi it)$ for $i = 0, 4; \ 1, 5; \ 2, 6; \ \text{and} \ 3, 7$.

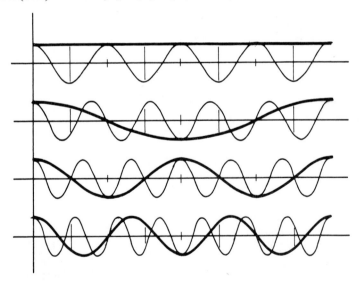

Fig. 3.1

To proceed, let us write (1) differently by expressing

$$i = \tfrac{1}{2}Ni_1 + i', \qquad j = 2j' + j_1 , \qquad (6)$$

where $j_1 = 0$ if j is even, $j_1 = 1$ if j is odd; $i_1 = 1$ if $i \geqslant \frac{1}{2}N$, $i_1 = 0$ if $i < \frac{1}{2}N$; and $0 < i', j' \leqslant \frac{1}{2}N-1$. We then have

$$\hat{X}_i = \sum_{j_1=0}^{1} \sum_{j'=0}^{\frac{1}{2}N-1} \exp[-c(\frac{1}{2}Ni_1+i')(2j'+j_1)/N] \, x[(2j'+j_1)/N]$$

$$= \sum_{j_1=0}^{1} \exp[-c(\frac{1}{2}Ni_1+i')j_1/N] \sum_{j'=0}^{\frac{1}{2}N-1} \exp(-ci'j'/\frac{1}{2}N) \exp(-ci_1j')$$

$$x[(2j'+j_1)/N] \, . \qquad (7)$$

The factor $\exp(-ci_1j')$ is always equal to 1. We are thus left with the fairly simple expression

$$\hat{X}_i = \sum_{j_1=0}^{1} \exp(-cij_1/N) \, x[i', j_1] \, , \qquad (8)$$

where

$$x[i', j_1] = \sum_{j'=0}^{\frac{1}{2}N-1} \exp(-ci'j'/\frac{1}{2}N) \, x[(2j'+j_1)/N] \, . \qquad (9)$$

(This is not a function of j' because we have already summed over all the possible values of j'.)

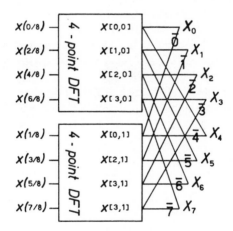

Fig. 3.2

Now $x[i', j_1]$ contains two parts. For $j_1 = 0$, the value $x[i', 0]$, $i' = 0$, 1, ..., $N-1$, are the N-point DFT of the even index values of x, $x(2j'/N)$; $x[i', 1]$ are the DFT of odd index values of x, $x[(2j'+1)/N]$. The two DFT's are combined by summing over j_1 as in (8). For example, for $N = 8$ $X_0 = x[0,0] + x[0,1]$ and $X_4 = x[0,0] + (-1)x[0,1]$, since both $i = 0$ and $i = 4$ give $i' = 0$; both $i = 1$ and $i = 5$ give $i' = 1$, so that $X_1 = x[1,0] + \bar{1} \cdot x[1,1]$ and $X_5 = x[1,0] + \bar{5} \cdot x[1,1]$. The whole computation process is shown in Fig. 3.2.

Another way of looking at the above process is in terms of matrix factorization. We have for $N = 8$

$$
\begin{vmatrix} \hat{X}_0 \\ \cdot \\ \cdot \\ \cdot \\ \cdot \\ \cdot \\ \cdot \\ \hat{X}_7 \end{vmatrix} = \begin{vmatrix} \bar{0} & 0 & 0 & 0 & \bar{0} & 0 & 0 & 0 \\ 0 & \bar{0} & 0 & 0 & 0 & \bar{1} & 0 & 0 \\ 0 & 0 & \bar{0} & 0 & 0 & 0 & \bar{2} & 0 \\ 0 & 0 & 0 & \bar{0} & 0 & 0 & 0 & \bar{3} \\ \bar{0} & 0 & 0 & 0 & \bar{4} & 0 & 0 & 0 \\ 0 & \bar{0} & 0 & 0 & 0 & \bar{5} & 0 & 0 \\ 0 & 0 & \bar{0} & 0 & 0 & 0 & \bar{6} & 0 \\ 0 & 0 & 0 & \bar{0} & 0 & 0 & 0 & \bar{7} \end{vmatrix} \begin{vmatrix} x[0,0] \\ x[1,0] \\ x[2,0] \\ x[3,0] \\ x[0,1] \\ x[1,1] \\ x[2,1] \\ x[3,1] \end{vmatrix} , \begin{vmatrix} x[0,0] \\ \cdot \\ \cdot \\ \cdot \\ \cdot \\ \cdot \\ \cdot \\ x[3,1] \end{vmatrix} = \begin{vmatrix} W_4 & & 0 \\ \hline & & \\ 0 & & W_4 \end{vmatrix} \begin{vmatrix} x(0) \\ x(2) \\ x(4) \\ x(6) \\ x(1) \\ x(3) \\ x(5) \\ x(7) \end{vmatrix} \quad (10)
$$

where W_4 is just $\exp(-cij/\tfrac{1}{2}N)$. More explicitly

$$
W_4 = \begin{vmatrix} \bar{0} & \bar{0} & \bar{0} & \bar{0} \\ \bar{0} & \bar{2} & \bar{4} & \bar{6} \\ \bar{0} & \bar{4} & \bar{0} & \bar{4} \\ \bar{0} & \bar{6} & \bar{4} & \bar{2} \end{vmatrix} .
$$

(Take care to distinguish $\bar{0}$ from 0; $\bar{0}$ is in fact $\exp(0) = 1$.) Note that we have grouped together even and odd index elements of x.

Notice that we have already reduced the required computation. To use the 'naive' method of (1), we must perform N additions and N multiplications for each i. For N values of i we have N^2 operations of each kind. We perform, however, two $\tfrac{1}{2}N$-point DFT's to obtain $x[i',0]$ and $x[i',1]$, using $2(\tfrac{1}{2}N)^2 = \tfrac{1}{2}N^2$ operations, followed by N multiplications and N additions, one for each $\hat{X}_i$. Altogether, only $N(\tfrac{1}{2}N+1)$ operations are needed. But we do not stop here. Each of the two $\tfrac{1}{2}N$-point DFT's can be reduced in the same manner into a combination of two $\tfrac{1}{4}N$-point DFT's, which can again be reduced. Thus,

$$
W_4 \begin{vmatrix} x(0) \\ x(2) \\ x(4) \\ x(6) \end{vmatrix} = \begin{vmatrix} \bar{0} & 0 & \bar{0} & 0 \\ 0 & \bar{0} & 0 & \bar{2} \\ \bar{0} & 0 & \bar{4} & 0 \\ 0 & \bar{0} & 0 & \bar{6} \end{vmatrix} \begin{vmatrix} W_2 & & 0 \\ \hline & & \\ 0 & & W_2 \end{vmatrix} \begin{vmatrix} x(0) \\ x(4) \\ x(2) \\ x(6) \end{vmatrix} ,
$$

where there is again a grouping of even-odd terms. Written out in full (1) becomes

$$
\begin{vmatrix} \hat{X}_0 \\ \cdot \\ \cdot \\ \cdot \\ \cdot \\ \cdot \\ \cdot \\ \hat{X}_7 \end{vmatrix} = \begin{vmatrix} 0 & 0 & 0 & 0 & \bar{0} & 0 & 0 & 0 \\ 0 & \bar{0} & 0 & 0 & 0 & \bar{1} & 0 & 0 \\ 0 & 0 & \bar{0} & 0 & 0 & 0 & \bar{2} & 0 \\ 0 & 0 & 0 & \bar{0} & 0 & 0 & 0 & \bar{3} \\ \bar{0} & 0 & 0 & 0 & \bar{4} & 0 & 0 & 0 \\ 0 & \bar{0} & 0 & 0 & 0 & \bar{5} & 0 & 0 \\ 0 & 0 & \bar{0} & 0 & 0 & 0 & \bar{6} & 0 \\ 0 & 0 & 0 & \bar{0} & 0 & 0 & 0 & \bar{7} \end{vmatrix} \begin{vmatrix} \bar{0} & 0 & \bar{0} & 0 & 0 & 0 & 0 & 0 \\ 0 & \bar{0} & 0 & \bar{2} & 0 & 0 & 0 & 0 \\ \bar{0} & 0 & \bar{4} & 0 & 0 & 0 & 0 & 0 \\ 0 & \bar{0} & 0 & \bar{6} & 0 & 0 & 0 & 0 \\ 0 & 0 & 0 & 0 & \bar{0} & 0 & \bar{0} & 0 \\ 0 & 0 & 0 & 0 & 0 & \bar{0} & 0 & \bar{2} \\ 0 & 0 & 0 & 0 & \bar{0} & 0 & \bar{4} & 0 \\ 0 & 0 & 0 & 0 & 0 & \bar{0} & 0 & \bar{6} \end{vmatrix} \begin{vmatrix} \bar{0} & \bar{0} & 0 & 0 & 0 & 0 & 0 & 0 \\ \bar{0} & \bar{4} & 0 & 0 & 0 & 0 & 0 & 0 \\ 0 & 0 & \bar{0} & \bar{0} & 0 & 0 & 0 & 0 \\ 0 & 0 & \bar{0} & \bar{4} & 0 & 0 & 0 & 0 \\ 0 & 0 & 0 & 0 & \bar{0} & \bar{0} & 0 & 0 \\ 0 & 0 & 0 & 0 & \bar{0} & \bar{4} & 0 & 0 \\ 0 & 0 & 0 & 0 & 0 & 0 & \bar{0} & \bar{0} \\ 0 & 0 & 0 & 0 & 0 & 0 & \bar{0} & \bar{4} \end{vmatrix} \begin{vmatrix} x(0) \\ x(4) \\ x(2) \\ x(6) \\ x(1) \\ x(5) \\ x(3) \\ x(7) \end{vmatrix} . \quad (11)
$$

This is shown in pictorial form in Fig. 3.3 where we have replaced $\overline{k+\tfrac{1}{2}N}$ by $-\bar{k}$ in accordance with (3). This is the Cooley-Tukey algorithm, the most common of the FFT algorithms. It is the only one that we study here.

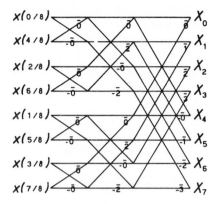

Fig. 3.3

Merits and de-merits of the FFT

Computation of an 8-point DFT as in (11) involves three steps, each requiring 8 multiplications and 8 additions. In general, the operation count for an N-point FFT is $N(\log_2 N)$, compared with N^2 if we use (1) directly. As an example of possible savings, when $N \sim 1000$ the FFT needs 100 times less effort than the naive method, and this ratio increases with N. But for FFT, it would have been quite impossible to Fourier transform any sizable amount of data. It has also been shown that FFT suffers less from roundoff errors. Thus, the merits of FFT are obvious. Its de-merits, however, are not as well known. Hence, we shall talk mostly about the latter, in order to persuade the reader not to use FFT blindly.

First of all, do not use FFT if you require only a small number of $\hat{X}$ values. An examination of Fig. 3.3 shows that it takes almost as much work to compute a single $\hat{X}$ as to obtain the lot. If we simply use (1), computing one $\hat{X}$ takes only N operations.

The second problem with FFT is that it cannot be performed in 'real time', in the sense that each $x(t)$ is being measured and brought into the computer separately as computation proceeds. With (1), we can just multiply $x(j/N)$ into $\exp(-cij/N)$ for every i and add the N products to N accumulators. We cannot do this in FFT. As Fig. 3.3 shows, in FFT we start off by combining $x(0)$ with $x(\tfrac{1}{2}N/N)$. The N input values are processed in whole to produce N other numbers, which are again processed together in the next stage. In short, we require all N values of x at the beginning of the FFT process.

For FFT to be possible, N must be a highly composite number. We have only really looked at the case when N is a power of 2, which is commonly assumed. It gives the largest possible savings in computing time. However, this is sometimes inconvenient, since often the length of input is fixed by other factors and may be somewhere between two powers of 2. We are then forced to add a string of zeros to the data in order to make the length a power of 2. This is equivalent to abruptly terminating the data, and gives rise to leakage. Consequently it is sometimes desirable to choose an N which contains non-binary factors, e.g., $N = 2^n 3^m$. The FFT algorithm for such values of N is somewhat more complex and also less efficient.

Another point here is that (1) can be computed entirely in real arithmetic. Thus,

$$\text{Re}(X_i) = \Sigma \cos(2\pi ij/N)\, x(j/N)$$

and

$$\text{Im}(X_i) = \Sigma \sin(2\pi ij/N)\, x(j/N)\,.$$

The situation is very different in FFT. We note that in (8) we multiply $\exp(cij_1/N)$ into $x[i', j_1]$. Similar products occur throughout the FFT process. In general, both multiplicands are complex, so that there are really *four* separate products here (R $\times$ R, R $\times$ I, I $\times$ R, I $\times$ I), whereas multiplying $\exp(cij/N)$ into $x(j/N)$ requires only two multiplications because the latter is real. In short, there is the hidden cost of complex arithmetic in FFT.

A not-so-hidden cost is that the vector $x(j/N)$ has to be sorted into a different order before computation is carried out. Recall that in (10) we collected together all the even index elements of x into the first half and the odd index ones into the second half. (Later, each half is divided into two quarters in the same way, and so on.) Now, an even integer j has 0 as its last binary digit, $j_1 = 0$; and odd integer has $j_1 = 1$. On the other hand, being in the first half of a vector means that the first binary digit of the index is 0; being in the second half requires a leading digit 1. In other words, the sorting goes as if we moved the last digit of j to the front:

$$j_n j_{n-1} \rightarrow j_1 j_n j_{n-1} \cdots j_2\,.$$

Thus $j_n \cdots j_2$ gives the order of a number within each half of x. Then in further simplifying (10) we did the same to each half of x. This implies that we keep j_1 unchanged, i.e., do not exchange elements between the two halves, but move j_2 to the front of the remaining digits. Thus, we now have $j_1 j_2 j_n \cdots j_3$. On continuing the sorting process we eventually get

$$j_n j_{n-1} \cdots j_1 \rightarrow j_1 j_2 \cdots j_n\,.$$

This is what we call the *bit reversal* operation. To illustrate, we again refer to Fig. 3.3, where x has been sorted into the order 0, 4 (100, or 001 reversed), 2(010), 6 (110), 1(001), etc. While the sorting process is trivial in theory, programming is quite a minor headache.

Finally, the FFT is more complex than the naive method. This produces the usual hidden cost of extra programming and de-bugging effort, additional compiling and execution time, etc.

Taking into account all these hidden costs, we find that FFT gives no real savings in computing time until N is well over 50. Even when N is somewhat over this the savings are not always large enough to make FFT mandatory. Obviously, if Fourier transformation is performed many times then the savings are worthwhile; otherwise, maybe not. There are also other circumstances, some we shall see in later chapters. In any case, it is advisable to consider all the alternatives and use a little common sense.

FFT of real data

As we noted in the last chapter, the N-point DFT of a real function is half redundant. Yet, if we use FFT we have to proceed as if we need all the results, since computing half of them takes no less time than computing all. This highly unsatisfactory state of affairs is neatly resolved by the method we describe below. The discussion is purely for the benefit of those readers who, for some reason or other, have to write their own FFT programs. Others can skip this section. It is worth mentioning that, some standard FFT subroutines do make use of the technique.

We have for real x:

$$\hat{X}_{N-i} = \Sigma \exp[-c(N-i)j/N]\, x(j/N) = \hat{X}_i^* \,, \qquad i = 1, 2, ..., \tfrac{1}{2}N-1. \quad (12)$$

It is easily seen that $\hat{X}_0$ and $\hat{X}_{\frac{1}{2}N}$ are purely real, so that the above is not relevant. Now let us suppose we have two sets of data, y and z, from which we are to compute $\hat{Y}$ and $\hat{Z}$. We construct a *complex* x by

$$x(j/N) = y(j/N) + \sqrt{-1}\, z(j/N) \,. \quad (13)$$

As x is no longer real the redundancy is eliminated. However, $\hat{Y}$ and $\hat{Z}$ are half redundant still. This permits us to recover $\hat{Y}$ and $\hat{Z}$ from $\hat{X}$. We have

$$\hat{X}_i = \hat{Y}_i + \sqrt{-1}\, \hat{Z}_i \,, \quad (14)$$

but

$$\hat{X}_{N-i} = \hat{Y}_{N-i} + \sqrt{-1}\, \hat{Z}_{N-i} = \hat{Y}_i^* + \sqrt{-1}\, \hat{Z}_i^* \,,$$

so that

$$\hat{X}_{N-i}^* = \hat{Y}_i - \sqrt{-1}\, \hat{Z}_i \,. \quad (15)$$

Combining (14) and (15) we have

$$\hat{Y}_i = \tfrac{1}{2}(\hat{X}_i + \hat{X}_{N-i}^*) \quad (16)$$

and

$$\hat{Z}_i = \tfrac{1}{2}\sqrt{-1}\,(\hat{X}_{N-i}^* - \hat{X}_i) \,. \quad (17)$$

These are valid for any i, including 0. (Remember aliasing, which makes $\hat{X}_N$ equal $\hat{X}_0$.) In particular, note that they give real results for $i = 0$ and $i = \tfrac{1}{2}N$.

We have just shown that: when our data are real, one DFT does the job of two! Given two real functions, we combine them using (13), take DFT of the complex values, and then 'unscramble' the two transforms by (16) and (17). Given two Fourier transforms of real functions, we combine them using (14), take Fourier series, whose real and imaginary parts then give y and z.

It is common to have two separate functions to transform in multivariate spectral analysis, but it does not always happen that way. However, even with only one set of real data, we can still take advantage of the savings provided by the method. In our derivation of FFT we first separated the even and odd index elements of $x(j/N)$, and computed a $\frac{1}{2}N$-DFT for each half, which we call $x[i', 0]$ and $x[i', 1]$. These are just $\hat{Y}$ and $\hat{Z}$. In other words, we put the even index elements into (13) as y and the odd index ones in as z, and obtain $x[i', 0]$ and $x[i', 1]$ from (16) and (17). We then use (8) to find $\hat{X}$. In this way, we Fourier transform N real numbers using only one $\frac{1}{2}N$-point FFT.

Recovery of the sampled values of x from $\hat{X}$ can also be simplified. First we must find $x[i', 0]$ and $x[i', 1]$ from $\hat{X}_{i'}$ and $\hat{X}_{i'+\frac{1}{2}N}$. Equation (8) gives

$$\hat{X}_{i'} = x[i', 0] + i' \cdot x[i', 1] \, , \quad \hat{X}_{i'+\frac{1}{2}N} = x[i', 0] - i' \cdot x[i', 1] \, .$$

Addition of the two equations gives

$$x[i', 0] = \frac{1}{2}(\hat{X}_{i'} + \hat{X}_{i'+\frac{1}{2}N}) \, , \tag{18}$$

and subtraction gives

$$x[i', 1] = \frac{1}{2}\exp(ci'/N)(\hat{X}_{i'} - \hat{X}_{i'+\frac{1}{2}N}) \, . \tag{19}$$

As $x[i', 0]$ is the DFT of even index values of x, we know that

$$x(2j'/N) = \frac{1}{N} \sum_{i'=0}^{\frac{1}{2}N-1} \exp(ci'j'/\frac{1}{2}N)(\hat{X}_{i'} + \hat{X}_{i'+\frac{1}{2}N}) \, ,$$

with a similar expression for $x[(2j'+1)/N]$. Here are the y and z vectors. That is, we put $(\hat{X}_{i'} + \hat{X}_{i'+\frac{1}{2}N})$ into (14) as $\hat{Y}_{i'}$, and $\exp(ci'/N)(\hat{X}_{i'} - \hat{X}_{i'+\frac{1}{2}N})$ as $\hat{Z}_i$, and compute the resulting complex Fourier series at $j' = 0, 1, ..., \frac{1}{2}N-1$. The real part of each result then gives $x(2j'/N)$, and the imaginary part, $x[(2j'+1)/N]$. Note that $\hat{X}_{i'+\frac{1}{2}N}$ is the same as $\hat{X}^*_{\frac{1}{2}N-i}$.

Sometimes we wish to Fourier transform two-dimensional data, for example pictures. This is done by Fourier transformation of each row, producing an equal number of transforms. These are then taken column by column and transformed a second time. The first time the data are real, so that we can combine two rows into one as in (13) and simplify things. The input for the second transform is no longer real, but it is still possible to obtain simplifications. However, this is beyond the scope of this book.

Forward and inverse transforms

So far in chapters 2 and 3 we have treated the transformation from x to X differently from the recovery of x from X. The former is a *forward* transformation, and the latter the *inverse* transformation. This distinction is real at first, when $x(t)$ has a continuous but finite argument, while X depends on an integer with an unlimited range. Once we introduce DFT, that distinction disappears. We have derived the FFT algorithm for the forward transformation, but in fact the algorithm can be used for both forward and inverse transforms. Let us see how this is done. The topic is simple, but it is also a potential source of confusion, which is why we give it a separate section.

We saw on page 16 that, for even N we recover the sampled values of x from the DFT $\hat{X}$ by the somewhat strange formula

$$x(j/N) = \sum_{i=-\frac{1}{2}N}^{\frac{1}{2}N} \frac{1}{2}\hat{X}_i \exp(cij/N) + \sum_{i=-\frac{1}{2}N+1}^{\frac{1}{2}N-1} \frac{1}{2}\hat{X}_i \exp(cij/N).$$

In the two parts the terms for $i = -\frac{1}{2}N+1, ..., \frac{1}{2}N-1$ are identical. Two end terms stand out, $\frac{1}{2}[\hat{X}_{\frac{1}{2}N} \exp(\frac{1}{2}cj) + \hat{X}_{-\frac{1}{2}N} \exp(-\frac{1}{2}cj)]$. Now, $\exp(\frac{1}{2}cj)$ and $\exp(-\frac{1}{2}cj)$ are the same because of periodocity, and $\hat{X}_{-\frac{1}{2}N} = \hat{X}_{\frac{1}{2}N}$ because of aliasing.† Hence, these two terms are really one, just $\hat{X}_{\frac{1}{2}N} \exp(c\frac{1}{2}Nj/N)$. Further, the terms for $i = -\frac{1}{2}N+1, ..., -1$ can be replaced by $i' = N+i = \frac{1}{2}N+1, ..., N-1$. Thus, we can write the above as

$$x(j/N) = \sum_{i=0}^{N-1} \hat{X}_i \exp(cij/N), \tag{20}$$

which is very similar to

$$\hat{X}_i = \sum_{j=0}^{N-1} x(j/N) \exp(-cij/N)/N. \tag{21}$$

The computation of (20) can be simplified in much the same way as that of (21). All the $\bar{k}$ terms that appear in (5), (10) or (11) are replaced by $-\bar{k}$, which is the same as $\overline{N-k}$ or $\bar{k}*$. Many FFT subroutines can be called with a parameter which specifies whether the caller wants $\exp(cij/N)$ or $\exp(-cij/N)$ used in the computation. (The FFT subroutine of Appendix 3 has this feature.)

However, if one has only a subroutine for (21), it is still possible to use it for (20). We have

$$x*(j/N) = \sum_{i=0}^{N-1} \hat{X}_i^* \exp(-cij/N). \tag{22}$$

---

† On a number of occasions we shall see this 'stand out' behaviour of $\frac{1}{2}N$. Another term that stands out is the zeroth. The reason lies in that neither has a distinct negative: $-0 = 0$ and $-\frac{1}{2}N$ is the same as $\frac{1}{2}N$ because of aliasing. Frequently, we divide these terms by 2 to avoid 'duplication'.

If x is real, then $x^* = x$ and $\hat{X}_i^* = \hat{X}_{-i} = \hat{X}_{N-i}$ so that

$$x(j/N) = \sum_{i=0}^{N-1} \hat{X}_{N-i} \exp(-cij/N) . \tag{23}$$

Thus, by some additional re-ordering we can use one subroutine for two purposes.

Let us see how this affects the simultaneous evaluation of two functions y and z using one FFT. Equation (14) becomes

$$\hat{X}_i = \hat{Y}_{N-i} + \sqrt{-1}\,\hat{Z}_{N-i} . \tag{24}$$

We perform a *forward* FFT on this $\hat{X}$ to recover y as the real part and z the imaginary part of x. Hence (18) and (19) become, upon replacement of i' by $\tfrac{1}{2}N-i'$:

$$x[i',0] = \tfrac{1}{2}(\hat{X}_{\frac{1}{2}N-i'} + \hat{X}_{N-i'}) , \tag{25}$$

$$x[i',1] = -\tfrac{1}{2}\exp(-ci'/N)(\hat{X}_{\frac{1}{2}N-i'} - \hat{X}_{N-i'}) . \tag{26}$$

That is, we put $(\hat{X}_{\frac{1}{2}N-i'} + \hat{X}_{N-i'})$ into (14) as $\hat{Y}_{i'}$ and $\exp(-i'/N)(\hat{X}_{N-i'} - \hat{X}_{\frac{1}{2}N-i'})$ into (14) as $\hat{Z}_{i'}$, to produce a complex vector whose real part gives $x(2j'/N)$ and imaginary part $x[(2j'+1)/N]$ after a *forward* FFT. Note that $\hat{X}_{N-i'}$ is the same as $\hat{X}_{i'}^*$. This fact is useful in that often we do not compute the second, redundant half of $\hat{X}$, so that (25) and (26) have to be evaluated using only the first half of $\hat{X}_i$, for $i = 0, 1, ..., \tfrac{1}{2}N$. The same remark applies to (18) and (19). (For example, see the subroutine FFTR in Appendix 2.)

4 Random Processes

Random variables

A *random variable* is something whose value cannot be predicted, except to the extent that it may take on one of a set of values, each with some probability. A *random process* is a function $x(t)$ whose value at any point in time cannot be predicted, except probabilistically, from its past values. Thus, the position of a point particle in a central field is not a random process, since we can compute the complete trajectory of the particle by measuring its position at two different moments. On the other hand, the Brownian motion of a micro-particle suspended in a liquid cannot be determined beforehand by computation. Similarly, the number of α-particles emitted by a grain of radioactive salt is a random process.

Actually, the line of demarcation between random and non-random variables is far from sharply defined. For example, it is in theory possible (classically speaking) to compute Brownian motion by first measuring the position and velocity of all the molecules around the micro-particle and then applying Newton's equations of motion to determine the changes in the motion of the particle caused by the bombardment of these molecules. In practice this is quite beyond our capabilities, so for all purposes the movement of the particle is random. To take a second example, if we are not over concerned with accuracy we can treat the motion of a rocket as non-random, since it is easy to approximately compute its trajectory from its initial speed, angle of departure, wind velocity, etc. However, if our wish is to pinpoint it precisely on some target, then we must take into account various unpredictable disturbances, which means that the speed, position, orientation, etc., must be treated as random variables.

Why should we be concerned with random processes in a discussion on spectral analysis? The motivation here is twofold. First, the signals we measure or receive are often contaminated by *noise*, which is a general name for unpredictable disturbances interfering with the signals. They come from numerous sources, and are of various degrees of adversity. Although noise is not exactly predictable, some of its behaviour on average may be discovered by observation over extended periods of time, and ways may be found to combat it. Second, the signals we measure or receive must also be unpredictable to some extent: if the future values of a function can be exactly predicted from its past, then there is no need to measure the future values at all! So again it is the behaviour of the signal on average that we find ourselves studying. In fact, communication engineers earn their living by designing systems which will, on average, remove as much as possible of unpredictable noise while preserving as much as possible of equally unpredictable signals!

While the past of a random process does not determine its value at time t, we may nevertheless know its *probability distribution*. Take, for example, the random draw from a bag containing M black balls and N white balls. If, say, we have already taken m black balls and n white balls out of the bag in the previous $(m+n)$ draws, this will not tell us in general the colour of the next ball drawn. However, it does show that there are $(M-m)$ black balls and $(N-n)$ white ones remaining in the bag, so that the chance of drawing a white ball next time is $(N-n)/(M+N-m-n)$. In a different case, while the number of α-particles emitted by a grain of salt is not exactly predictable, its probability distribution is a known function of time. (It is a Poisson distribution with a parameter that decreases exponentially with time.)

This brings us to the concept of *a priori* and *a posteriori* probabilities. Let us again look at the example of ball drawings. Suppose we started with a bag containing 2 white and 2 black balls, and have already taken one ball out but did not look at its colour. What is the chance of getting a white ball next draw? Since we have not seen the drawn ball, there is chance ½ that there are now 2 white balls and 1 black ball left in the bag, and also ½ that there are 2 black balls and 1 white ball. The former means that there is chance ⅔ of drawing white next time; in the latter case the chance is ⅓. The chance of drawing black are exactly the reverse. It is clear that, in the absence of any knowledge about the previously drawn ball, we must say that we are equally likely to draw either white or black. On the other hand, if we already know that the previous draw is black, then the chance for drawing a white ball next time must be ⅔. This is the *a posteriori* probability, the probability for some event after we know about some other event. The *a priori* probability, in this case ½, is the probability for some event when we do not have knowledge about some other event.

(Beginning students often have trouble understanding the concepts just discussed, and sometimes ask: 'Since the process of drawing balls is independent of our thoughts, how can our knowing or not knowing about the previous draw affect the result of the next draw?' The answer is, it does not affect the result of the draw itself. But it does affect our ability to *predict about* the next draw. It is important to realize that, probability has more to do with our ability to predict the behaviour of something than the behaviour itself. Anything we do not know about exactly before actually measuring it can be considered as random, which is why probability theory is so often applied to problems which do not seem to have any randomness in their nature.)

Let $P(A)$ and $P(B)$ be the *a priori* probabilities of events A and B, and $P(B/A)$ the *a posteriori* probability for B knowing that A has actually occurred. If we know that A cannot possibly affect B, or, though A may affect B we do not know enough about their relation, then knowing whether A has occurred does not help us to predict B. In this case there is no difference between the *a priori* and *a posteriori* probabilities, and $P(B) = P(B/A)$. Such two events A and B are said to be independent. For example, if after each draw we return the drawn ball to the bag, then the result of the previous draws have no effect on future draws. A somewhat different case is the following: suppose our bag contains 10^{100} balls of each colour, then regardless of whether we have drawn white or black the first time the chance for getting either colour next time is still for all purposes ½. In this case knowing the colour of the first ball drawn has no practical influence on our capacity to predict the next draw, even though in theory the influence is there.

Now we introduce the concept of *joint probability*, $P(A, B)$, the chance of both A and B occurring. This is an *a priori* probability because we have not assumed any knowledge of other relevant events. We shall state without proof that

$$P(A, B) = P(A) P(B/A) = P(B) P(A/B) . \qquad (1)$$

The relation can be looked at this way: In the absence of any other knowledge, A has chance $P(A)$ of occurring. Now knowing that A has actually occurred, the chance of B occurring is $P(B/A)$. The chance of both occurring must increase with each of the two probabilities, and obviously (1) satisfies this requirement. It is clear from (1) that, if A and B are independent then

$$P(A, B) = P(A) P(B) . \qquad (2)$$

It is important to remember that the above relation is not generally valid except if A and B are independent. For example, choose A and B to be the same event. Then, if A occurs B must also occur, so $P(B) = P(A)$. On the other hand, $P(A, B)$ is also just $P(A)$ since A occurring is no different from both A and B occurring. Clearly, $P(A, B) \neq P(A) P(B)$ in general.

Values of P must always be positive or zero. Negative probabilities have no sensible meaning. It is usual to normalize P by making the sum of the probabilities of all possible events equal 1. When something is certain to occur, so that it is the only possible event, its probability is then 1. The reverse is not necessarily true. An example will make this clear. When we throw a dart at a board, its chance of hitting a particular point is zero, since a point is supposed to have zero area. But this does not mean that the dart cannot hit that point. Similarly, the chance of *not* hitting a particular point is $(1-0) = 1$, but we may nevertheless hit that point. Consequently, we can only say that an event with probability 1 is *almost* certain to occur. The 'almost' is essential.

Probability distributions and averages

We are now ready to go into more quantitative discussions of probability. Suppose x can take any one of a set of values x^i, $i = 1, 2, ..., M$. The set is called the *ensemble* of x. (We shall not discuss in detail the case of x being able to take any value within a continuous range as this is mathematically more troublesome. However, provided M is large enough and the x^i are spaced closely enough we can approximate a continuous range to any desirable accuracy. Further, we should remember that our measuring instruments cannot return arbitrary accuracy, and we can compute with finite digit numbers only, so it is in fact more realistic to assume discrete values. However, a continuous ensemble is very convenient to use, as shown in the next section.) From this we can define a set of M events, E_i, being the event of x actually taking value x^i, and a set of probabilities $P(E_i)$. Note that

$$\sum_{i=1}^{M} P(E_i) = 1 ,$$ (3)

since x has to fall somewhere among the M values. It is clearer to write $P(E_i)$ as $P_x(x^i)$. Thus, $P_x(a)$ is the probability of x taking the value a, assuming that a is among the M values. For a pair of random variables x and y, we write the joint probability of x taking the value x^i and y taking the value y^j as $P_{xy}(x^i, y^j)$.

Now let us suppose that x is a function of time, $x = x(t)$. As t changes, x takes one of the M values but with changing probability for each value, depending on what the past values have been. Thus, $P_x(x^i)$ is a function of time, and it also depends *a posteriori* on what we know about the past values. We write as $P_x^t(x^i)$ the *a priori* probability that x takes the value x^i at time t. This is the probability when we know nothing about the past values. We write as $P_x^{t,s}(x^i, x^j)$ the probability that x takes the value x^i at time t and the values x^j at time s. This is again an *a priori* probability as no other knowledge is assumed here. The *a posteriori* probability of x_j taking value x^i at t knowing that it has taken value x^j at s is of course just $P_x^{t,s}(x^i, x^j)/P_x^s(x^j)$. Other, more complicated types of probabilities may be defined but are not required for our purposes.

It is useful to know that

$$\sum_{j=1}^{M} P_x^{t,s}(x^i, x^j) = P_x^t(x^i) ,$$ (4)

since the left hand side is the total probability of x taking value x^i at time t regardless of what value x takes at s. If $x(t)$ is such that knowing it at time s does not improve our ability to predict its value at t, then

$$P_x^{t,s}(x^i, x^j) = P_x^t(x^i) P_x^t(x^j) . \tag{5}$$

We also note that, if we define a new random function $z(x)$ in terms of x, then z can take one of the M values $z^i = z(x^i)$, $i = 1, 2, ..., M$. Further, the probability for each value of z is the same as the probability of x taking the corresponding value, or,

$$P_x^t(x^i) = P_z^t(z^i) ,$$

assuming that the correspondence between z and x is unique. If, however, $z(x)$ is such that more than one value of x^i give the same value $z(x)$, then the probability of z taking that value is the sum of all the probabilities of the corresponding values of x.

A random process is said to be *time invariant*, or *stationary**, if

$$P_x^t(x^i) \text{ does not vary with } t,$$

and

$$P_x^{t,s}(x^i) \text{ varies with } (t-s) \text{ only.} \tag{7}$$

The second property means that, no matter what t is, knowing $x(t-\Delta)$, its value Δ time units earlier, has the same effect on our ability to predict $x(t)$, since $P_x^{t,t-\Delta}$ varies only with Δ, not with t.

We define the *ensemble average* (or simply *average*) of random variable x as

$$\langle x \rangle = \sum_{i=1}^{M} P_x(x^i) x^i . \tag{8}$$

Applying the same to $z(x)$ we get

$$\langle z \rangle = \sum_{i=1}^{M} P_z(z^i) z^i = \sum_i P_x(x^i) z(x^i) . \tag{9}$$

As $\langle x \rangle$ and $\langle z \rangle$ may vary with time, except that if x is time invariant then P_x^t does not change with t so that $\langle x \rangle$ and $\langle z \rangle$ would be constant. Hence $\langle x \rangle$ is also called the *mean* of x.

Let us choose $z(x) = x^2$. Its average is called the *mean square* of x:

$$\langle x^2 \rangle = \sum_{i=1}^{M} P_x(x^i)(x^i)^2 . \tag{10}$$

The mean square of a random variable indicates how significant x is, numerically speaking, on average. Thus, $\langle x^2 \rangle = 0$ implies that x is identically zero, since every term in the summation of (10) is non-negative so that the sum can be zero only when every term is zero. In view of (3), at least some of the probabilities must be non-zero, which means that they multiply into zeros in (10). In other words, x has zero probability of being non-zero. In contrast, $\langle x \rangle$ being zero would not in general mean that x is identically zero.

---

* The correct term is '*wide sense* stationary (time invariant)'. A stationary process may may be regarded as having a stable underlying structure.

Yet another important average is the *variance*

$$V(x) = \langle (x - \langle x \rangle)^2 \rangle,$$

which measures the deviation of x from its own mean value. If $V(x) = 0$, then x is constant since it does not depart from its mean. Variance is also called *mean square deviation* or *dispersion*. $[V(x)]^{1/2}$ is called the *standard deviation*, being in a sense the average amount by which x deviates from $\langle x \rangle$, and is usually denoted as σ_x. A useful relation is

$$V(x) = \langle x^2 \rangle - 2\langle x \langle x \rangle \rangle + \langle \langle x \rangle^2 \rangle = \langle x^2 \rangle - \langle x \rangle^2 . \tag{11}$$

Given two random variables x and y, with joint probability distribution $P_{xy}(x^i, y^j)$, we define their *covariance* as

$$C(x, y) = \langle (x - \langle x \rangle)(y - \langle y \rangle) \rangle . \tag{12}$$

More explicitly this is

$$C(x, y) = \langle xy \rangle - \langle x \langle y \rangle \rangle - \langle y \langle x \rangle \rangle + \langle x \rangle \langle y \rangle = \langle xy \rangle - \langle x \rangle \langle y \rangle,$$

with

$$\langle xy \rangle = \sum_{i,j=1}^{M} P_{xy}(x^i, y^j) x^i y^j .$$

An important relation is that

$$\langle x^2 \rangle \langle y^2 \rangle \geqslant \langle xy \rangle^2 . \tag{13}$$

To show this, let us form new random variables $\xi = x/\langle x^2 \rangle^{1/2}$ and $\eta = y/\langle y^2 \rangle^{1/2}$. Clearly, $\langle \xi^2 \rangle = \langle \eta^2 \rangle = 1$. Now

$$0 \leqslant \langle (\xi + \eta)^2 \rangle = \langle \xi^2 \rangle + \langle \eta^2 \rangle + 2\langle \xi\eta \rangle,$$

and

$$0 \leqslant \langle (\xi - \eta)^2 \rangle = \langle \xi^2 \rangle + \langle \eta^2 \rangle - 2\langle \xi\eta \rangle,$$

which gives

$$1 \geqslant \langle \xi\eta \rangle \geqslant -1, \qquad \text{or,} \qquad \langle xy \rangle^2 / \langle x^2 \rangle \langle y^2 \rangle \leqslant 1 .$$

This proves (13). We also have $\langle x^2 \rangle + \langle y^2 \rangle \geqslant 2\langle xy \rangle$, proved similarly.

Let us suppose that we are now looking at two random variables which are values of random process x at times t and s. Their covariance

$$\langle x(t) x(s) \rangle = \sum_{i,j} P_x^{t,s}(x^i, x^j) x^i x^j, \tag{14}$$

is a function of t and s. This is called the *autocorrelation* function* of x. If x is time invariant this, like $P_x^{t,s}$, varies with $(t - s)$ only. This allows us to write

$$\langle x(t) x(s) \rangle = a(t - s) . \tag{15}$$

Further, as $\langle x(t) x(s) \rangle$ and $\langle x(s) x(t) \rangle$ are exactly the same thing, $a(t - s) = a(s - t) = a[-(t - s)]$. In other words, a is an even function, and we can write it as $a(|t - s|)$ if we so wish.

*This is the usual engineering text definition. Statisticians insist that this is sloppy practice. We prefer to follow the trend.

If $t = s$, then $a(t-s) = a(0)$ is just the mean square of $x(t)$. Note that this is independent of t, as x is time invariant. It is also non-negative. Further, in view of (13) we have

$$a(0)^2 = \langle x(t)^2 \rangle \langle x(s)^2 \rangle \geqslant \langle x(t)\, x(s) \rangle^2 \,,$$

or

$$a(0) \geqslant |a(t)| \,, \qquad t \neq 0 \,. \tag{16}$$

When $x(t)$ is an electric signal, its mean square gives its average *power*. This then is the physical significance of $a(0)$. For non-zero time differences $a(t)$ shows the relation between the values of x at different times. We shall see that, in linear signal processing systems the autocorrelation function of a random process provides us with essentially all the information required to handle it mathematically.

The autocorrelation function of a random process can be computed if we know about P_x. Unfortunately, this is seldom the case. There are random signals for which we can with some confidence construct a mathematical model, e.g., thermal noise, but for most signals we can only try to estimate a from measured values by averaging $x(s)\, s(t)$ over different times. Mathematically, however, this time average is not necessarily the same as average over ensemble. They are equivalent only when the signal is *ergodic*. The question is then how we can show that a random process is in fact ergodic. This is a very difficult question, but it can be said that we cannot establish ergodicity, nor time invariance, nor randomness, by measuring the values of a signal. Again we must appeal to what we know about its origin and be persuaded that there is no reason to believe that it is not ergodic. It is well for the reader to know that, although statistical communication has had many successes its theoretical foundation is still not without shakiness.

Some random processes*

A random variable is known as *Gaussian* if its ensemble ranges from $-\infty$ to $+\infty$ at constant intervals, i.e., $x^i = i\Delta$, $i = \dots, -1, 0, 1, \dots$, and

$$P_x(x^i) = \frac{\Delta}{\sqrt{2\pi}\, \sigma_x} \exp(-\tfrac{1}{2} i^2 \Delta^2 / \sigma_x^2) \,, \tag{17}$$

where σ_x is the standard deviation of x and acts as an independent parameter. If σ_x is small, then only values close to 0 have significant probabilities.

Let us assume that Δ is so small compared with attainable accuracy of measurement that, for all practical purposes the range appears to be concinuous. This assumption is important in that, when we wish to take the average of $z(x)$ we can replace the summation

$$\sum_{i=-\infty}^{\infty} P_x(x^i)\, z(x^i) = \frac{1}{\sqrt{2\pi}\, \sigma_x} \sum_{i=-\infty}^{\infty} \exp(-\tfrac{1}{2} i^2 \Delta^2 / \sigma_x^2)\, \Delta z(i\Delta) \,,$$

---

* Those who do not like hairy mathematics should browse through this section rather quickly.

by the integral
$$\frac{1}{\sqrt{2\pi}\,\sigma_x} \int_{-\infty}^{\infty} \exp(-\tfrac{1}{2}x^2/\sigma_x^2)\, z(x)\, dx \, , \qquad (18)$$

upon letting $i\Delta = x$ and $dx = \Delta$. The importance of Gaussian random variables is partly due to the ease with which integrals of the form of (18) can be manipulated, and partly due to the fact that, when a random variable is generated by the combination of a large number of unrelated random disturbances it is approximately Gaussian. Brownian motion, for example, falls in this class because it is caused by the bombardment of liquid molecules into the micro-particle. Thus, Gaussian variables are useful as well as mathematically easy to handle. The distribution is shown in Fig. 4.1.

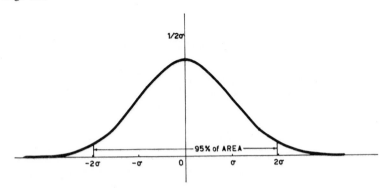

Fig. 4.1

It is clear that
$$\langle x \rangle = \frac{1}{\sqrt{2\pi}\,\sigma_x} \int_{-\infty}^{\infty} \exp(-\tfrac{1}{2}x^2/\sigma_x^2)\, x\, dx = 0 \, . \qquad (19)$$

Other equations, shown in many textbooks, are
$$\frac{1}{\sqrt{2\pi}\,\sigma_x} \int_{-\infty}^{\infty} \exp(-\tfrac{1}{2}x^2/\sigma_x^2)\, dx = 1 \, , \qquad (20)$$

$$\frac{1}{\sqrt{2\pi}\,\sigma_x} \int_{-\infty}^{\infty} \exp(-\tfrac{1}{2}x^2/\sigma_x^2)\, x^2\, dx = \langle x^2 \rangle = \sigma_x^2 \, , \qquad (21)$$

and more generally,
$$\langle x^n \rangle = 0 \, , \quad n \text{ odd}; \qquad = (n-1)(n-3)\ldots 1 \cdot \sigma_x^n \, , \quad n \text{ even}. \qquad (22)$$

Another useful relation for Gaussian variables is
$$\langle xyvw \rangle = \langle xy \rangle\langle vw \rangle + \langle xv \rangle\langle yw \rangle + \langle xw \rangle\langle yv \rangle \, . \qquad (23)$$

Let us study the new random variable x^+ produced from x by adding constant μ. We have
$$\langle x^+ \rangle = \langle x \rangle + \mu = \mu \, ,$$

while
$$\langle (x^+)^2 \rangle = \langle x^2 \rangle + 2\mu \langle x \rangle + \mu^2 = \sigma_x^2 + \mu^2 .$$

However,
$$V(x^+) = \langle (x^+ - \langle x^+ \rangle)^2 \rangle = \langle (x^+)^2 \rangle - \langle x^+ \rangle^2 = \sigma_x^2 .$$

Now, we can consider $\langle x^2 \rangle$ to be the variance of x since its mean is 0. This shows that x and x^+ have the same variance but different means, or, adding a constant to a Gaussian variable shifts the mean but does not change standard deviation.

(The point to remember here is that, the mean of a random variable is by itself often quite meaningless. Thus, yearly average temperature of a city tells us very little about what clothes to take if moving there, since the summer may be very hot and winter very cold even though the mean may be quite reasonable. Even daily averages may belie the severe nights and steaming days. To ensure comfort, we need to know the daily, monthly and yearly fluctuations in the temperature, the deviations. Recall also that, $\langle x^2 \rangle = 0$ implies x is always 0, while $\langle x \rangle = 0$ would not be sufficient.)

On substituting $x = x^+ - \mu$ into (18) we see that x^+ has the probability distribution

$$P_{x^+} = \frac{1}{\sqrt{2\pi}\,\sigma_x} \exp[-\tfrac{1}{2}(x^+ - \mu)^2/\sigma_x^2)\, dx ; \qquad (24)$$

x^+ is also said to be Gaussian variable. It is then clear that we can always reduce a Gaussian variable to one with zero mean simply by subtracting its mean, without changing the standard deviation. In fact, it is common to assume that we always subtract the mean of random variables so that we can confine our discussion to zero mean variables.

We also see that $x' = ax$ is a Gaussian variable, and has probability distribution

$$P_{x'} = \frac{1}{\sqrt{2\pi}\,\sigma_x} \exp(-\tfrac{1}{2}x'^2/a^2\,\sigma_x^2)\, d(x'/a) , \qquad (25)$$

which is as if σ_x has been replaced by $a\sigma_x$. This implies that x' has standard deviation $a\sigma_x$. Thus, by adding a constant or multiplying by a factor we can always change a Gaussian random variable to one with any desirable mean and variance.

Now let us study the sum of two *independent* zero mean Gaussian variables $z = x + y$. The joint probability distribution of x and y is (for simplicity we assume $\sigma_x = \sigma_y = \sigma$):

$$P_{xy} = \frac{1}{\sqrt{2\pi}\,\sigma^2} \exp[-\tfrac{1}{2}(x^2 + y^2)/\sigma^2]\, dx\, dy . \qquad (26)$$

To find the distribution of z, we add up (i.e., integrate) the probabilities of those x and y such that $y = z - x$, which gives the integral

$$P_z = \frac{dy}{2\pi\sigma^2} \int_\infty^\infty \exp[-\tfrac{1}{2}(x^2/\sigma^2) - \tfrac{1}{2}(z-x)^2/\sigma^2]\, dx$$

$$= \frac{dy}{2\pi\sigma^2} \int_\infty^\infty \exp[-\tfrac{1}{2}(\sqrt{2}x - z/\sqrt{2})^2/\sigma^2 - \tfrac{1}{2}(z/\sqrt{2})^2/\sigma^2]\, dx .$$

Changing variable of integration to $\sqrt{2}\,x - z/\sqrt{2}$ we get

$$P_z = \frac{dy}{\sqrt{2\pi}\,\sigma}\,\exp[-\tfrac{1}{2}z^2/(\sqrt{2}\,\sigma)^2]\,\frac{1}{\sqrt{2\pi}\,\sigma}\int\exp[-\tfrac{1}{2}(x')^2/\sigma^2]\,dx'/\sqrt{2}\;.$$

And, making use of (20) we get simply

$$P_z = \frac{1}{\sqrt{2\pi}\,(\sqrt{2}\,\sigma)}\,\exp[-\tfrac{1}{2}z^2/(\sqrt{2}\,\sigma)^2]\,dz\;,$$

since $dz = dy$. Thus, z is a Gaussian variable with mean 0 and variance $2\sigma^2$. It is easy to show that the sum of N independent Gaussian variables has variance $N\sigma^2$. Further, even if the N variables are not independent the sum is still Gaussian, though the variance is not so easy to find. (For example, if $z = x + x$, then $\langle z^2 \rangle = 4\sigma^2$, not $2\sigma^2$.) In general, any linear combination of Gaussian variables is a Gaussian variable.

Since the joint distribution of two independent variables is just the product of their respective distributions, we have

$$\langle xy \rangle = \langle x \rangle \langle y \rangle\,, \tag{27}$$

or, their covariance, $\langle xy \rangle - \langle x \rangle \langle y \rangle$, is zero. If two random variables, independent or otherwise, satisfy (27), they are said to be *uncorrelated*. While independence implies uncorrelatedness, the reverse is true only if the variables are Gaussian. Further the joint distribution of two non-independent Gaussian variables can be written down quite compactly, even though it is not simply the product of their separate distributions. We shall state without proof that, for Gaussian variables x and y with standard deviations σ_x and σ_y and zero means,

$$P_{xy} = \frac{1}{2\pi\sigma_x\,\sigma_y\sqrt{1-\rho^2}}\,\exp[(-\tfrac{1}{2}x^2/\sigma_x^2 - \tfrac{1}{2}y^2/\sigma_y^2 + \rho xy/\sigma_x\,\sigma_y)/(1-\rho^2)]\,dx\,dy\,, \tag{28}$$

where

$$\rho = \langle xy \rangle/\sigma_x\,\sigma_y\,,$$

is called the correlation coefficient between x and y. (Derivation of this can be found in Davenport and Root. For non-zero mean variables replace x by $x - \mu_x$ and y by $y - \mu_y$.)

Since we know the *a priori* probability distribution P_y, we can also find the *a posteriori* probability distribution $P_{x/y}$, the probability for x taking some value when the value of y is known, by the simple relation

$$P_{x/y} = P_{xy}/P_y = \frac{1}{\sqrt{2\pi}\,\sigma_x\sqrt{1-\rho^2}}\,\exp[(-\tfrac{1}{2}x^2/\sigma_x^2 + \tfrac{1}{2}\rho^2\,y^2/\sigma_y^2 + \rho xy/\sigma_x\,\sigma_y)/(1-\rho^2)]$$

$$= \frac{1}{\sqrt{2\pi}\,(\sigma_x\sqrt{1-\rho^2})}\,\exp[-\tfrac{1}{2}(x - \sigma_x\,\rho y/\sigma_y)^2/\sigma_x^2(1-\rho^2)]\;. \tag{29}$$

This describes a Gaussian variable with mean $\sigma_x\,\rho y/\rho_y$ and variance $\sigma_x^2(1-\rho^2)$. Or, once we know y we are better able to predict x, since it now has a smaller standard deviation, reduced by a factor of $(1-\rho^2)^{1/2}$, and would be distributed about the mean of $\sigma_x \rho y / \sigma_y$ rather than 0. If, for example, $\rho = 1$, then x has zero variance, which means it is constant, and has mean $\sigma_x \rho y / \sigma_y$, to which it is always equal. In other words, if correlation coefficient is 1 then measuring one variable automatically fixes the other also. Note that, because of (13) ρ^2 cannot exceed 1.

A Gaussian random process is one whose value at any point in time is Gaussian. Thus, we require $P_x^t(x^i)$ to be of the form of (17). But in addition, the joint distribution of any set of its values at different points in time is also restricted to be of Gaussian form. In particular, the joint distribution of two values must be of the form of (28).

An example of random processes is the following; given a set of independent random variables $z(t)$, with zero mean, standard deviation σ, and $z(t)$ uncorrelated with $z(t')$ if $t \neq t'$, we take the initial value of new process $x(t)$, $x(0)$, to be the first z, $z(0)$, and then generate $x(t)$ from $x(t-\Delta)$ by the formula

$$x(t) = \alpha x(t-\Delta) + (1-\alpha^2)^{1/2} z(t) .$$

Since $x(t-\Delta)$ include only those $z(t')$ with $t' < t$, $x(t-\Delta)$ is uncorrelated with $z(t)$. We assume that Δ is small enough so that the process $x(t)$ appears to change continuously with time. It is clear that

$$\langle x(t) \rangle = 0 , \qquad \langle x(t)^2 \rangle = [\alpha^2 + (1-\alpha^2)] \sigma^2 = \sigma^2 .$$

Thus, x has the same mean and mean square as z for this choice of α. Also, it is easy to find its correlation function. We have that, for $t = s + i\Delta$

$$x(s+i\Delta) = \alpha x[s+(i-1)\Delta] + (1-\alpha^2)^{1/2} z(s+i\Delta)$$
$$= \alpha^2 x[s+(i-2)\Delta] + \alpha(1-\alpha^2)^{1/2} z[s+(i-1)\Delta] + \alpha(1-\alpha^2)^{1/2} z(s+i\Delta)$$
$$= \ldots = \alpha^i x(s) + \text{terms uncorrelated with } x(s).$$

This means

$$\langle x(s+i\Delta) x(s) \rangle = \alpha^i \langle x(s)^2 \rangle = \alpha^{(t-s)/\Delta} \sigma^2 .$$

We can re-write this as

$$\langle x(t) x(s) \rangle = A \exp[B(t-s)] ,$$

where A and B are constants depending on a, σ and Δ. This is for $t \geqslant s$. For $s < t$ we obviously have

$$\langle x(t) x(s) \rangle = A \exp[B(s-t)] ,$$

so that, in general

$$\langle x(t) x(s) \rangle = A \exp(B |t-s|) . \tag{30}$$

(Note that B must be negative, as $|\alpha|$ is required to be less than 1.)

A special case of the above is $\alpha = 0$, or, $x = z$, with the additional requirement that z be Gaussian. This gives $\langle x(t)^2 \rangle = \sigma^2 \neq 0$ while

$\langle x(t) x(s) \rangle = 0$ if $t \neq s$. Or,

$$\langle x(t) x(s) \rangle = \Omega \delta (t-s) . \tag{31}$$

Here Ω is a constant and δ is the δ-function, something which is zero everywhere except where its argument is zero, i.e., at $t = s$. It is thus a 'spike'. The δ-function is the functional analog of the Kronecker delta. It has the property

$$\int \delta(t-s) f(s) \, ds = f(t) ,$$

since $\delta(t-s) = 0$ except at $s = t$, so that only $f(t)$ contributes to the integral. Implicit here is the assumption that the range of integration includes the point $s = t$. Equation (31) says that values of x at different times are completely independent. One value tells us nothing about those at other times. This particular random process is called *white noise*. (It is to be remembered that white noise must be Gaussian, even though many authors define white noise by (31) alone. As we shall see later, white noise is not a physically producible signal, but is only a mathematical abstraction.)

Let us now study a different random process, $z(t)$, with a finite ensemble z_i, $i = 1, 2, ..., M$. Now $z(t)$ has the characteristic that it has probability $p \, dt$ of making a jump into any of its permissible values during any time period t to $t + dt$. There is thus probability $[1-(M-1)p \, dt]$ that it will *not* make a transition, remaining at the former value. Thus,

$$P(z = z_i \text{ at } t + dt) = P(z = z_i \text{ at } t)[1-(M-1)p \, dt] + P(z \neq z_i \text{ at } t)p \, dt .$$

Since $P(z \neq z_i \text{ at } t) = 1 - P(z = z_i \text{ at } t)$, we have

$$P(t + dt) = P(t) - (M-1) p P(t) \, dt + p[1 - P(t)] \, dt ,$$

where we denote $P(z = z_i \text{ at } t)$ by simply $P(t)$. The above is just

$$P'(t) = \frac{P(t + dt) - P(t)}{dt} = -MpP(t) + p .$$

Let us suppose that at time s, z is known to be equal to z_i. This allows us to impose the initial condition $P(s) = 1$. $P(t)$ is the *a posteriori* probability that $z = z_i$ at t knowing $z = z_i$ at s. Solving the differential equation we get

$$P(t) = \frac{1}{M} + \frac{M-1}{M} \exp[-pM(t-s)] .$$

However, *a priori* all levels are equally likely, thus

$$P_z^s(z_i) = 1/M ,$$

which gives

$$P_z^{t,s}(z_i, z_i) = P(t) P_z^s(z_i) = P(t)/M ,$$

while

$$P_z^{t,s}(z_{j \neq i}, z_i) = [1 - P(t)] (M-1)^{-1} P_z^s(z_i)$$

$$= M^{-2} \{ 1 - \exp[-pM(t-s)] \} .$$

The mean of z is just $\Sigma_i\, z_i/M$, while its mean square is $\Sigma_i\, z_i^2/M$. Now we can derive $\langle z(t)\, z(s)\rangle$. First we make use of $P_z^{t,s}(z_i, z_j)$ for $i = j$ and $i \neq j$. Then we note that $\Sigma_{i \neq j}$ is the same as $\Sigma_{i,j} - \Sigma_{i=j}$, so that we have the following long string of derivations:

$$\langle z(t)\, z(s)\rangle = \sum_{i,j} z_i\, z_j\, p_z^{t,s}(z_j, z_i)$$

$$= \sum_i (z_i^2/M)\left(\frac{1}{M} + \frac{M-1}{M}\exp[-pM(t-s)]\right)$$

$$+ \sum_{i \neq j} (z_i\, z_j/M^2)\{1 - \exp[-pM(t-s)]\}$$

$$= \langle z^2\rangle\left(\frac{1}{M} + \frac{M-1}{M}\exp[-pM(t-s)]\right) + \sum_i \sum_j \frac{z_i}{M}\frac{z_i}{M}\{1 - \exp[-pM(t-s)]\}$$

$$- \sum_i (z_i^2/M^2)\{1 - \exp[-pM(t-s)]\}$$

$$= \langle z^2\rangle\exp[-pM(t-s)] + \langle z\rangle^2\{1 - \exp[-pM(t-s)]\}$$

$$= \langle z^2\rangle\exp[-pM(t-s)]\ .$$

We have so far assumed $t \geqslant s$. Consideration of the case for $t < s$ easily leads to the general result

$$\langle z(t)\, z(s)\rangle = \langle z^2\rangle\exp(-pM\,|t-s|)\ . \tag{32}$$

We have shown that, although z here has a different model it has the same autocorrelation function as the process considered earlier. In general, very different random processes can appear to be similar because they have approximately the same autocorrelation. The fact gives a warning that, when one is mathematically modelling a physical signal even though the autocorrelation given by the model closely resembles that of the signal we cannot conclude that the model must be a good one. However, for many practical purposes it does happen that autocorrelation alone matters.

For a stationary random process, once we know $a(t)$ and $\langle x(t)\rangle$ we also know the covariance between any two values of x as well as their correlation coefficient, since by definition

$$C[x(t), x(s)] = \langle x(t)\, x(s)\rangle - \langle x(t)\rangle\langle x(s)\rangle$$

$$= a(t-s) - \mu^2\ .$$

(Remember that x has a constant mean.) The correlation coefficient is

$$\rho[x(t), x(s)] = \langle [x(t)-\mu]\,[x(s)-\mu]\rangle/\sigma_{x(t)}\,\sigma_{x(s)}$$

$$= [a(t-s) - \mu^2]/\sigma_{x(t)}^2\ .$$

(Again x has a constant standard deviation too.) But

$$\sigma_{x(t)}^2 = V[x(t)] = \langle [x(t)-\mu]^2\rangle = \langle x(t)^2\rangle - \mu^2 = a(0) - \mu^2\ .$$

Thus, the correlation coefficient between $x(t)$ and $x(s)$ is just the value $[a(t-s) -\mu^2]/[a(0) -\mu^2]$. For a Gaussian process, once we know μ, σ and ρ we can immediately write down the joint probability distribution of $x(t)$ and $x(s)$ by the use of (28). In short, the autocorrelation function $a(t)$ summarizes a lot of information about the random process. It has a special significance in spectral analysis as it is the Fourier transform of the power spectrum.

Autoregressive processes

An important class of random processes is as follows:

$$x(t) = \sum_{j=1}^{m} \alpha_j\, x(t-j\Delta) +z(t)\,, \tag{33}$$

where z is white noise. If $m = 1$, this is just the first example of random processes we saw in the previous section. The expression (33) is called an mth order *autoregressive process* by statisticians, since x depends on its own past values. The dependence, however, is mixed with unpredictable disturbances provided by the values of z. Autoregressive processes provide reasonably good models of certain real life systems. For example, the population of a species of animal within a region certainly depends on its past values, but it is also affected by random factors like the weather. It is also reasonable to think that present values of stocks determine the trading pattern in the immediate future and therefore future prices, with random factors again playing a part.

Let us consider the autocorrelation of x. First we have

$$\langle x(t)^2 \rangle = \sum_{j=1}^{m} \alpha_j \langle x(t)\,x(t-j\Delta)\rangle +\langle x(t)\,z(t)\rangle\,, \tag{34}$$

where $z(t)$ is completely independent of the previous values of x as it is an unpredictable disturbance. Therefore, by the use of (33) we have

$$\langle x(t)\,z(t)\rangle = \langle z(t)^2 \rangle = \sigma^2\,,$$

so that (34) becomes

$$a(0) = \sum_{j=1}^{m} \alpha_j\, a(j\Delta) +\sigma^2\,. \tag{35}$$

We then look at $\langle x(t)\,x(t-k\Delta)\rangle$ for $k \geqslant 1$. We know $\langle x(t-k\Delta)\,z(t)\rangle = 0$, and this gives

$$a(k\Delta) = \sum_{j=1}^{m} \alpha_j\, a(|k-j|\Delta)\,. \tag{36}$$

There is one more relation. Correlating each side of (33) with itself gives, again remembering that $z(t)$ is independent of past values of x:

$$a(0) = \sum_{j,j'=1}^{m} \alpha_j\, \alpha_{j'}\, a(|j-j'|\Delta) +\sigma^2\,. \tag{37}$$

Suppose we have been given an autoregressive model, i.e., given m and α_j for $j = 1, 2, ..., m$, and we wish to compute $a(t)$. By combining (37) with (36) for $k = 1, 2, ..., m-1$, we have m equations in m unknowns, $a(0), ..., a[(m-1)\Delta]$. This specifies the first m values of $a(t)$. After solving the equations, we can then use (36) to find $a(k\Delta)$ for $k = m, m+1, ... $.

As an example, we consider the case of $m = 1$, where (37) gives

$$a(0) = \alpha_1^2 \, a(0) + \sigma^2 ,$$

so that

$$a(0) = \sigma^2/(1-\alpha_1^2) . \tag{38}$$

Equation (36) gives

$$a(k\Delta) = \alpha_1 \, a[(k-1)\Delta] . \tag{39}$$

Applying (39) repeatedly we have

$$a(k\Delta) = \alpha_1^k \, a(0) = \alpha_1^k \, \sigma^2/(1-\alpha_1^2) .$$

This of course agrees with the result of the last section. It should be pointed out that $|\alpha_1|$ must be restricted to less than 1; otherwise the process is not even stationary, since we would have $a(k\Delta) > a(0)$.

It is harder for $m = 2$, when (37) gives

$$a(0) = (\alpha_1^2 + \alpha_2^2) \, a(0) + 2\alpha_1 \, \alpha_2 \, a(\Delta) + \sigma^2 , \tag{40}$$

and (36) for $k = 1$ is

$$a(\Delta) = \alpha_1 \, a(0) + \alpha_2 \, a(\Delta) . \tag{41}$$

Values for $a(0)$ and $a(\Delta)$, and thereafter $a(k\Delta)$ for general k, can be obtained with no difficulty, though the general expressions are rather messy. Things get even worse for larger m, but we shall not be studying these.

In practice the opposite problem, that of finding the model, m and α, from measured values of $a(t)$, is more frequent. For this, we combine (36) with (35) and solve for the α's. Since, however, we usually do not know what m is initially, the general practice is to start with $m = 1$, find α_1, then go to $m = 2, 3$, etc., until the α's no longer change significantly. For details see Box and Jenkins, chapter 3, or Kendal, chapter 12 for a briefer discussion.

Once we have measured the autocorrelation function, we can compute the spectrum from it. Statisticians, however, often favour an indirect method: fit an autoregressive model to the process by computing α's from $a(t)$, and then compute the spectrum from the α's. (Chapter 10, though formally on a different subject, would show the reader how to do the latter if he is really keen to know.) The method has both merits and shortcomings. Most unfortunately, it is rather laborious to use. We shall not discuss it any further.

Non-stationary processes

When we apply the theory of stationary random processes to signal processing, the implicit assumption is of course that signals found in real life can be modelled in such ways with reasonable accuracy. Quite often, we are not really sure whether this is true, but as long as the models are not manifestly wrong we might just go ahead and use them, regardless of philosophical niceties. There are times, however, when the signals we obtain are clearly non-stationary. That is, certain portions of the signal show clearly different properties from other parts. This can take the form of a *trend*, either a steady drift in the mean value of the signal as it is measured, or perhaps a gradual change in the degree of fluctuation, which corresponds to a trend in the variance rather than the mean. It is also possible to have a periodic structure superimposed on random fluctuations. Other, more complex types of non-stationarity are possible.

Trends and periodicities should be removed if they appear to dominate the other features; otherwise the spectrum we compute would mainly be that of the non-stationarity. The techniques for doing this are rather crude, since we are really just taking a guess at what the non-stationarity is. The more we know about the nature of the signal and its measurement process, the better we would be able to make that guess, which makes the actual procedure rather application dependent. We shall therefore just take a brief look at a few common techniques.

Perhaps the most common trend is a linear drift, which occurs if our measuring apparatus gets out of calibration gradually, or if the signal has been subjected to integration, which converts a constant mean into a linear one. Even when the trend is not just a straight line, if it is slow enough then simply removing the linear part is often quite sufficient for eliminating most of the effect the trend causes. To remove a linear trend, we first find a linear function that best approximates the data, and then subtract the function from each data point. Equations for finding the function are given by (2.2), except that now our $y_i(t)$ is a power of t: $y_0 = 1$, $y_1 = t$, and $y_i = t^i$ in general. Thus, we wish to fit to $x(t)$ the polynomial

$$\sum_{i=0}^{m} \alpha_i t^i .$$

(42)

This requires us to solve the equations

$$\sum_j A_{ij} \alpha_j = \int t^i x(t) \, \mathrm{d}t ,$$

(43)

with

$$A_{ij} = \int t^i t^j \, \mathrm{d}t = \int t^{i+j} \, \mathrm{d}t .$$

When fitting a straight line, we take $m = 1$, which produces the equations

$$\alpha_0 + \tfrac{1}{2}\alpha_1 = \int x(t) \, \mathrm{d}t \qquad \text{and} \qquad \tfrac{1}{2}\alpha_0 + \tfrac{1}{3}\alpha_1 = \int x(t) t \, \mathrm{d}t ,$$

which amount to

$$\alpha_0 = 4 \int x(t) \, \mathrm{d}t - 6 \int x(t) t \, \mathrm{d}t \qquad \text{and} \qquad \alpha_1 = 12 \int x(t) t \, \mathrm{d}t - 6 \int x(t) \, \mathrm{d}t .$$

If the trend appears to be more complicated than linear, we might try fitting a polynomial of degree 2 or 3. This requires the solution of 3 or 4 equations in as many unknowns. Fitting even higher order polynomials is normally of little use. The reason for this is briefly discussed in section Appendix 2.

Periodicities in the data are quite a different matter. They are seldom produced by instrumental faults, though this does occur, e.g., when the output is mixed with interference from the AC power supply. Mostly, however, periodicities are due to 'seasonal factors' or 'natural rhythm' in the producing system. Consequently, they can be extremely important indicators of the system's structure, and we should not simply eliminate them. Instead, we should separate the two parts of the signal and analyse each part. This can be achieved by the use of digital filtering, which is discussed in chapter 11, though there are many other methods. Occasionally, trends can also be physically significant. When we are uncertain about the cause of the trend, it may be advisable to separate it out and analyse it to see if it does contain useful information.

Once we are reasonably sure that the time variation in the mean has been eliminated we can then take steps to remove the time dependence in the variance. Doing so requires the multiplication of the data by a function inversely proportional to the square root of the variance. However, multiplication is the same as addition of logarithms. Thus, the easiest thing to do is to take the logarithm of the absolute value of each value of x, converting a multiplicative time variation to an additive one. This is then removed by the methods already discussed. It should be pointed out, however, that removal of variance trends disturbs the structure of the signal much more than mean trend removal. We should not attempt it unless we are very sure about its necessity.

If the signal is purely positive, then it is meaningful to perform spectral analysis on its logarithm. This is called cepstral analysis. It separates the signal into components that contribute in a *multiplicative* rather than additive manner. We discuss this type of analysis briefly in chapter 12.

When the non-stationarity in $x(t)$ is very complicated, there is no reliable method for removing it. Instead, we should divide the signal into pieces which appear to be reasonably time invariant, perhaps remove their trends, compute the spectrum for each piece, and then analyse each spectrum individually to see the short term behaviour of the signal, and compare different spectra to determine the long term behaviour. Analysis of a complex non-stationary signal as one piece is not usually a fruitful operation, as it amounts to trying to force the signal into an inappropriate stationary framework.

5 The Power Spectrum

Definition

Given the function $x(t)$, we can define other quantities in terms of it. Given a random process $x(t)$, we can do the same except that these quantities, like x, would be random in nature. They would take on a set of values with some probability for each value, and we can define their mean, mean squares, covariances, etc. The statistical properties of x then determine those of the derived quantities.

In spectral analysis, our purpose is to study the importance of certain kinds of regular structure in the behaviour of $x(t)$. When $x(t)$ is just one function, then we need only to Fourier transform it and then examine the relative sizes of the Fourier coefficients. This process is called *harmonic analysis*. When $x(t)$ is a random process, harmonic analysis is not very meaningful. For, as $x(t)$ is supposed to fluctuate randomly, the function we have measured is nothing more than one out of many possible functions, and its Fourier coefficients are what they are by chance. That one term happens to be large does not necessarily mean it is important *on average*. Thus, our job is to find quantities that indicate the importance of the Fourier components in an average sense. Fortunately, we need to know very little about the probabilistic structure of $x(t)$ or X_i in order to analyse them meaningfully.

Let us look at the Fourier coefficients of $x(t)$, a zero mean process:

$$X_i = \int_0^1 \exp(-cit)\, x(t)\, dt .$$

What are the mean and the mean square of X_i? Clearly,

$$\langle X_i \rangle = \int \exp(-cit) \langle x(t) \rangle\, dt = 0 , \tag{1}$$

and

$$\langle |X_i|^2 \rangle = \langle X_i^* X_i \rangle = \langle \int \exp(cit) x(t) \, dt \exp(-cis) x(s) \, ds \rangle$$

$$= \int \exp[-ci(s-t)] \langle x(t) x(s) \rangle \, ds \, dt . \tag{2}$$

Equation (2) gives the relation between the mean square of X_i and the auto-correlation function $a(t-s)$. It is clear from (1) that the mean of X_i is not a useful quantity. It is always 0 and tells us nothing about X_i. The mean square, however, is useful. We know

$$x(t) = \sum_i X_i \exp(cit) ,$$

so that the magnitude of X_i shows how important the ith term is for representing $x(t)$ by means of Fourier series. A different way of saying this is that, the power of $x(t)$ is $x(t)^2$ while the power of ith term is

$$| X_i \exp(cit) |^2 = | X_i |^2 .$$

Thus, $|X_i|^2$ is the power residing in the ith component. Parseval's theorem (p. 8) gives

$$\langle \int_0^1 x(t)^2 \, dt \rangle = \int a(0) \, dt = a(0) = \sum_i \langle |X_i|^2 \rangle . \tag{3}$$

Thus, the total signal power contained in $x(t)$ is equal to the sum of the powers of its Fourier components. To put it differently, the mean squares of X_i specify the distribution of the power of $x(t)$ among its Fourier components. This is why they are said to constitute the *power spectrum* of $x(t)$. We shall denote $\langle |X_i|^2 \rangle$ as S_i. And $|X_i|^2$, without the averaging, is denoted as Ξ_i. This is known as the *periodogram*.

Equation (2) relates S to $a(t)$ in a rather complex way. We shall now simplify this. Putting $s' = (s-t)$ we have [remember $a(t) = a(-t)$]:

$$S_i = \int_0^1 dt \int_{-t}^{1-t} \exp(-cis') a(s') \, ds'$$

$$= \int_0^1 dt \left(\int_{-t}^0 + \int_0^{1-t} \right) \exp(-cis') a(s') \, ds' . \tag{4}$$

In the first part, $-t \leqslant s' < 0$. Since $0 \leqslant t \leqslant 1$, we have $-1 \leqslant -t \leqslant s' \leqslant 0$, which is the same as $1 \geqslant t \geqslant -s' = |s'|$ and $-1 \leqslant s' \leqslant 0$. Thus, we convert the first integral into the following form by changing the order of integration:

$$\int_{-1}^0 ds \exp(-cis') a(s') \int_{|s'|}^1 dt = \int_{-1}^0 \exp(-cis') a(s')(1-|s'|) \, ds' .$$

In the second integral $0 \leqslant s' \leqslant 1-t \leqslant 1$. This is the same as $0 \leqslant s' \leqslant 1$ and $0 \leqslant t \leqslant 1-|s'|$. Hence we write the second integral as

$$\int_0^1 ds' \exp(-cis') a(s') \int_0^{1-|s'|} dt = \int_0^1 \exp(-cis') a(s')(1-|s'|) \, ds' .$$

Combining the two parts gives

$$S_i = \int_{-1}^{1} \exp(-cit)\, a(t)(1 - |t|)\, dt. \tag{5}$$

(Fig. 5.1 gives an intuitive picture of the transition from (2) to (5). Instead of integrating in the s and t directions, we could integrate along the two diagonals. However, $(t-s)$ is constant along the marked diagonals, and since the integrand is a function of $(t-s)$ only, it is also constant. Integration along these lines just give the length of the diagonal, which decreases linearly like $(1 - |t|)$. We then integrate along the other diagonal. Thus, we have the original integrand times a linearly decreasing function. Note the separation into two parts just divides the area of integration into two triangles.)

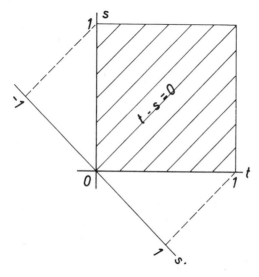

Fig. 5.1

We note that this is a Fourier transformation defined over $[-1, 1]$. It is, however, possible to obtain S_i by Fourier transformation over $[0, 1]$ if we so wish, since (5) is the same as

$$S_i = \int_{0}^{1} [\exp(-cit) + \exp(cit)]\, a(t)(1 - t)\, dt$$

$$= 2 \int_{0}^{1} \cos(2\pi it)\, a(t)(1 - t)\, dt = 2\mathrm{Re}(A_i), \tag{6}$$

where

$$A_i = \int_{0}^{1} \exp(-cit)\, a(t)(1 - t)\, dt, \tag{7}$$

i.e., A is the Fourier transform of $a(t)(1-t)$ over $[0, 1]$.

S_i is what $|X_i|^2$ should be on average, while $a(t)$ is what $x(s)\,x(s+t)$ should be on average. For this reason they are sometimes called *expectation values*. When we measure X_i, or for that matter any random variable, the values we get do not necessarily equal what it should be on average. It is thus important to distinguish between the *theoretical model* and the *measured values.* * We cannot compute S_i unless we are given the model. Given measured values of x, we can only make guesses about $a(t)$ or S might be. This is why spectral analysis is never a cut and dry business, since it is far from easy to know what guesses are good. For example, Ξ the periodogram, is on average equal to S by definition. Yet, we shall later show that Ξ is certainly not a good approximation to S because on average it deviates from S by as much as S.

Nevertheless, the periodogram is an important quantity, and we shall study it further. We have

$$\Xi_i = \int_0^1 \exp(cit)\,\exp(-cis)\,x(t)\,x(s)\,dt\,ds\ .$$

Performing the same manipulation as that preceding (5) we get

$$\Xi_i = \int_{-1}^1 \exp(-cit)\,\xi_a(t)\,dt\ , \tag{8}$$

where

$$\xi_a(t) = \int_0^{1-|t|} x(s+|t|)\,x(s)\,ds\ .$$

Because $\xi_a(t)$ is a time average of $x(s)\,x(s+t)$, we might expect it to be a good approximation to $a(t)$. We are now closer to the truth, as we shall show later. Equation (8) shows then that the periodogram is the Fourier transform of an approximation to $a(t)$ defined over the interval $[-1, 1]$.

Later on we shall see that we do not actually compute Ξ_i as our spectrum. Rather we compute

$$\Xi_i^w = \int_{-1}^1 \exp(-cit)\,w(t)\,\xi_a(t)\,dt\ , \tag{9}$$

where $w(t)$ is a *window* function with the following properties:

$$w(0) = 1, \quad w(-t) = w(t) \qquad \text{and} \qquad w(-t) = w(t) = 0, \quad T \leqslant t \leqslant 1,$$

for some chosen value T. This restricts the range of integration in (9) to $[-T, T]$. We also have in analogy to (6) and (7)

$$\Xi_i^w = 2\int_0^T \cos(2\pi it)\,w(t)\,\xi_a(t)\,dt = 2\mathrm{Re}\!\left(\int_0^T \exp(-cit)\,w(t)\,\xi_a(t)\,dt\right) . \tag{10}$$

---

* An obvious example is probability distribution versus histogram.

Some properties of Fourier transforms over $[-T, T]$

Over $[0, 1]$ we define the Fourier transform of $x(t)$ as

$$X_i = \int_0^1 \exp(-cit)\, x(t)\, dt, \qquad i = 0, \pm 1, \pm 2, \dots , \qquad (11)$$

and this set of numbers contain the same information as $x(t)$, since $x(t)$ is completely specified once we know X (assuming of course that $x(t)$ is continuous and finite),

$$x(t) = \sum_i X_i \exp(cit). \qquad (12)$$

One might think that, since S_i is uniquely determined by X_i and is related to $a(t)$ by an expression similar to (11), it should completely specify $a(t)$. This is *not* so. Rather, $a(t)$ is completely determined by A_i, which, however, contains a real and an imaginary part. The real part is S_i. The imaginary part cannot in general be computed from the real part. In other words, S_i, $i = 0, \pm 1, \pm 2, \dots$, contain less information than $a(t)$. And the same remark applies to Ξ_i in relation to $\xi_a(t)$.

The reason for this is quite simple. Equation (5) shows that S_i is the Fourier transform of $a(t)$ *defined over* $[-1, 1]$. This is an interval of length 2, and by the results on page 21 we should be taking the following:

$$S_{i/2} = \int_{-1}^1 \exp(-cit/2)\, a(t)\, dt. \qquad (13)$$

In other words, instead of using only exponential functions of integer frequencies, we should also take those with half integer frequencies. Only then do we recover all the information contained in $a(t)$. (Note, however, there is no simple relation between S for half integer frequencies and $\mathrm{Im}(A)$ for integer frequencies.)

Similarly, when we compute Ξ^w using a window that cuts the range of integration off at $t = \pm T$, all the information in $w(t)\,\xi_a(t)$ is recovered if we take

$$\Xi^w_{i/2T} = \int_{-T}^T \exp(-cit/2T)\, w(t)\, \xi_a(t)\, dt/2T. \qquad (14)$$

The exponentials used have frequencies $i/2T$. This in general is a non-integer. Consequently, most authors prefer to denote this as f, hence $\Xi^w(f)$, $\exp(-cft)$, etc. We choose not to follow this because identifying X, S, Ξ, etc., by i clearly brings out the fact that we compute transform at a discrete set of frequencies only. There is no confusion as long as the reader keeps in mind that i refers to the ith exponential function we need, not necessarily the function with frequency i. Rather, it has frequency i/L, L being the length of the interval over which integration takes place. However, whenever there is any possibility of confusion we shall write out the frequency explicitly, as in (13) and (14).

Since $X_{-i} = X_i^*$ we have

$$\Xi_{-i} = |X_{-i}|^2 = \Xi_i. \qquad (15)$$

(We remind the reader that i might be replaced by $i/2$ or $i/2T$ depending on the context!) Similarly, $S_{-i} = S_i$ and $\Xi_i^w = \Xi_{-i}^w$. These are also evident from (5), (8) and (14), because each of these is the Fourier transform of an even integrand. (I.e., we may replace t by $-t$, which results in the negation of i.)

Let us consider the computation of Ξ^w from sampled values. Given N sampled values of $x(t)$, we are able to obtain N values of $\xi_a(t)$ at $t = j/N$. However, we retain only those with $-T < t < T$. Thus, j is restricted to the range $-TN < j < TN$. Let us write TN as M. Equation (14) now becomes

$$\Xi_{i/2T}^w = \sum_{j=-M+1}^{M-1} \exp[-ci(j/N)/(2M/N)]\, w(j/N)\, \xi_a(j/N)/2M .$$

Combining the jth term with the $(-j)$th term, with the exception of $j = 0$ which does not have a negative counterpart, we get

$$\Xi_{i/2T}^w = [w(0)\,\xi_a(0) + 2 \sum_{j=1}^{M-1} \cos(2\pi ij/2M)\, w(j/N)\, \xi_a(j/N)]/2M$$

$$= 2\mathrm{Re}\left(\sum_{j=1}^{M-1} \exp(cij/2M)\, w(j/N)\, \xi_a'(j/N)/2M \right), \qquad (16)$$

where $\xi_a' = \xi_a$ except that $\xi'(0) = \xi_a(0)/2$. Equation (16) shows how we actually compute the spectrum from autocorrelation. The summation can be recognized as a $2M$-point discrete Fourier transform, except that there are only M terms in the sum. Thus, this is the discrete transform of a $2M$-element vector whose first M elements are $w(j/N)\,\xi_a'(j/N)$, while the last M elements are all zero. This addition of zeros, due to the necessity of using frequencies $i/2T$ rather than i/T, will also appear later in the computation of $\xi_a(t)$ by fast Fourier transform methods.

As on page 21, we have the very pleasant feature that, when we perform a $2M$-point DFT on ξ_a', we automatically have the correct frequencies $i/2T$ in the exponential function. We only have to hand the M values of ξ_a' and M zeros to a standard DFT subroutine. The spectrum we obtain in return would have a set of values, each corresponding to the power around frequency $i/2T$.

The situation remains equally pleasant even when our signal $x(t)$ has been measured over an interval of length L rather than the standard interval $[0, 1]$. Now the truncated autocorrelation is of length $2TL$, for $-TL < t < TL$. Therefore, we should use the frequencies $1/2TL$. Again, we need not be concerned about this during computation, i.e., we simply perform a $2M$-point DFT. Each value in the resulting spectrum can afterwards be identified with a frequency $i/2TL$.

The reader may have heard that, given M measured values we can only produce M spectral values at most, but here we are taking $2M$-point discrete Fourier transform and so would get $2M$ values of Ξ. Is there something funny here? The answer is no. We saw that, in a M-point transform, because of aliasing, the coefficients for $i = \frac{1}{2}M+1, \frac{1}{2}M+2, ..., M-1$ are really for $i = -\frac{1}{2}M+1, ..., -1$. But since $\Xi_i = \Xi_{-i}$, these values are redundant. We have $(M+1)$ values from $2M$-point transform, but only M of these are independent because the total power is fixed. If we take only M frequencies, we would in fact lose half of the numbers we should be getting.

The power spectra of simple random processes

We saw on page 44 that white noise is a zero-mean Gaussian process with the autocorrelation $\langle x(t)x(s) \rangle = \delta(t-s)$, which is 0 everywhere except at $t = s$. Its power spectrum is then

$$S_i = \Omega \int_{-1}^{1} \exp(-cit)\,\delta(t)(1-|t|)\,dt .$$

$\delta(t)$ is everywhere zero except at $t = 0$, so that

$$\langle |X_i|^2 \rangle = S_i = \Omega .$$

Thus, the power of white noise is equally distributed among its Fourier components. (In fact, it is equally distributed among any orthogonal set of functions.) We also have that the total power is infinite if we add up the power in all the components, from $i = -\infty$ to $+\infty$. This is why white noise is a physically unrealizable process. It is used as a mathematical approximation for those random processes which fluctuate very rapidly with no apparent regularity of any kind.

Let us study other properties of the Fourier coefficients. We have

$$\langle X_i^* X_j \rangle = \Omega \int_0^1 \exp(cit - cjs)\,\delta(t-s)\,dt\,ds .$$

Since $\delta(t-s)$ is zero everywhere but at $t = s$, we get

$$\langle X_i^* X_j \rangle = \Omega \int_0^1 \exp[c(i-j)t]\,dt = 0 \qquad \text{if } i \neq j .$$

Thus, different Fourier coefficients are uncorrelated. As $x(t)$ is Gaussian, so is X, in which case uncorrelatedness implies independence. Note that the above applies even to X_i and X_{-i} though they are *arithmetically* related as $X_i = X_{-i}^*$. This distinction between arithmetic and statistical independence is important, as we shall see shortly.

When we use sampled values of x, from N measured values we obtain N values of X, which we have shown to be independent Gaussian variables with zero mean and variance Ω. To be more precise, we should note that X has real and imaginary parts, each of which, being a linear combination of the values of a Gaussian process, is a Gaussian variable. Although $\langle X_i^* X_{-i} \rangle = 0$, the real and imaginary parts of X_i are not by themselves uncorrelated with those of X_{-i}. In fact, $\text{Re}(X_i) = \text{Re}(X_{-i})$ and $\text{Im}(X_i) = -\text{Im}(X_{-i})$

as shown on page 7. When we produce $\Xi_i = |X_i|^2$, $\Xi_i = \Xi_{-i}$, so that we get only $\tfrac{1}{2}N+1$ independent variables, these being Ξ_1 to $\Xi_{\frac{1}{2}N-1}$ plus Ξ_0 and $\Xi_{\frac{1}{2}N}$, the last two having no negative index counterparts. The Ξ values are not Gaussian as they are non-linear functions of Gaussian variables. Rather, they are what one calls χ^2 variables. Ξ for $i = 1$ to $\tfrac{1}{2}N-1$ contain two independent variables each, Re(X) and Im(X). They are said to have two degrees of freedom; Ξ_0 and $\Xi_{\frac{1}{2}N}$ have one degree only.

As we said before, to be able to recover all the information contained in $\xi_a(t)$ we need also to find X at half integer frequencies and then compute the corresponding Ξ. These values are not statistically independent of the integer frequency Ξ. Arithmetically, however, they are. There is no way of computing the half-integer frequency Ξ values from the integer frequency Ξ without further information. (It is possible to do this for the X values, but this is not the same thing because in general knowing $|X|^2$ is not enough to determine X.) This distinction between statistical independence and arithmetic independence is a source of much confusion, and there have been arguments about whether it is correct to evaluate Ξ at non-integer frequencies because 'they are dependent'. We certainly think it necessary to include half-integer frequencies. In fact, we see no reason why one should not evaluate X or Ξ at even small frequency intervals if this facilitates spectrum interpretation by making the features of a spectrum clearer. It is puzzling to us why some authors should be so against computing numbers which provide information already contained in other numbers. The information may be 'in there somewhere', but if it is very hard to see then it is of limited value. A little more computing time is not all that expensive these days.

When people compute the periodogram, they usually just perform an N-point DFT on x, which gives only $\tfrac{1}{2}N$ values of Ξ, the other half being redundant. We are of the opinion that one should perform a $2N$-point DFT, and obtain an N-point periodogram, with both integer and half-integer frequencies represented, up to the maximum of $\tfrac{1}{2}N$, which is the limit imposed by the sampling theorem. The N values can then be taken in groups and averaged to produce a good spectrum. We shall discuss this again later.

When we generate random numbers, we wish to have uncorrelated output, i.e., white noise. Fourier transform provides one, out of many, ways for testing their randomness. If a set of numbers is truly random, then it must have a constant power spectrum. However, this is a necessary but not sufficient condition. Further, although Ξ is on average supposed to equal Ω, it actually fluctuates a great deal about this mean value. Thus, the job is not as simple as it may sound.

When x is not white noise, S is not constant. Different values in the periodogram are not independent. The correlation between different periodograms, while not zero, tends to be much smaller than that between successive values of $\xi_a(t)$. This is why the spectrum tends to be easier to interpret.

We might say that each value in the spectrum represents a separate piece of information. In the autocorrelation much information is the interrelation between values. The function of the spectrum is then to pick out some of the interrelations. Also, because it is the interrelations rather than the numbers themselves that are important, we do little harm in multiplying $\xi_a(t)$ by $w(t)$ prior to Fourier transformation. This changes the values but makes little difference as far as the interrelations are concerned.

Two random processes we discussed in the previous chapter have the auto-correlation, shown in the upper part of Fig. 5.2

$$a(t) = A \exp(-B|t|).$$

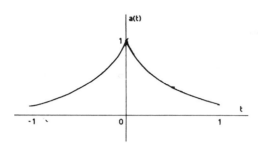

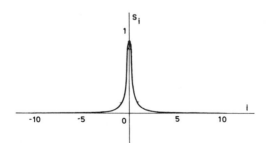

Fig. 5.2

Its spectrum S can be approximated quite easily. We have

$$S_i = \int_{-1}^{1} \exp(-cit)\, a(t)(1-|t|)\, dt.$$

We note that multiplication by $(1-|t|)$ has almost no effect on $a(t)$. Thus

$$S_i \sim A \int_0^1 \exp[-(ci+B)t]\, dt + A \int_{-1}^0 \exp[-(ci-B)t]\, dt$$

$$= A[1-\exp(-ci-B)]/(ci+B) + A[\exp(ci-B)-1]/(ci-B).$$

For integer i, $\exp(ci) = 1$; for half integer i, $\exp(ci) = -1$. Thus, in one case we have

$$S_i = 2B[1-\exp(-B)]/[B^2 + 4\pi^2 i^2],$$

while for half-integer frequencies

$$S_i = 2B[1 + \exp(-B)]/[B^2 + \pi^2 i^2] .$$

If B is large, then both have the shape of the lower portion of Fig. 5.2.

Note that when B is very large, $a(t)$ drops rapidly to 0, so that it comes close to being a $\delta(t)$, i.e., 0 everywhere except if $t = 0$. At the same time, S comes close to being constant. In other words, if B is large we again have white noise. On the other hand, if B is very small we have a spectrum that is concentrated near $i = 0$. In other words, x varies slowly and contains mainly low frequency terms.

On cross spectrum, etc.*

So far we have only looked at the spectral analysis of a random process standing alone. In numerous applications we are also interested in the relation between two random processes, and an important indicator of this relation is their *cross spectrum* and various quantities derived from it. With two processes, the mathematics of the estimation process become even more involved, but here we sketch the main ideas as a starting point for those readers who wish to know about the subject.

Let us take random processes $x(t)$ and $y(t)$. They may be simultaneous outputs of a physical system, or one may be the input signal that controls the generation of the other. In the second case, the output is a function of the input with the possible addition of some unpredictable interference. In the first case, x and y are both functions of some unknown stimulus. By studying the relation between x and y, we hope to understand the physical system to a point when something useful can be done with this knowledge, e.g., we may wish to compute y from measured values of x, or we may want to find out how the unknown stimulus is connected with x and y. Without going into specifics, let us say that the cross spectrum plays a part in this.

We have defined $\langle |X_i|^2 \rangle$ as the power spectrum of x and denoted it as S_i. Now that we have two processes, we have to distinguish between them. Thus, $\langle |X_i|^2 \rangle$ is now S_i^x and $\langle |Y_i|^2 \rangle$ is written as S_i^y, naturally. Further, we can now define the cross spectrum between x and y as

$$S_i^{xy} = \langle X_i Y_i^* \rangle = \int \exp[ci(s-t)] \langle x(t) y(s) \rangle \, dt \, ds$$

$$= \int_{-1}^{1} g(t)(1 - |t|) \exp(-cit) \, dt . \tag{17}$$

Here

$$g(t) = \langle x(s+t) y(t) \rangle .$$

Because S^{xy} is the Fourier transform of $g(t)(1 - |t|)$ over $[-1, 1]$ we shall also write it as G. Note that the properties of g are rather different from those of a. For example, $g(t)$ and $g(-t)$ are not necessarily equal. G_i is

* May be omitted on a first reading.

not necessarily real, nor symmetric. It has a real and an imaginary part, which are respectively called the cospectrum and quadspectrum. G_i can also be expressed in the polar form $G_i = |G_i| \exp(\sqrt{-1}\,\phi_i)$; $|G_i|$ is called the amplitude spectrum, and ϕ_i the phase spectrum. These all have their particular uses in applications. Yet another derived quantity is the coherency spectrum:

$$\kappa_i = |G_i|^2/(S_i^x S_i^y)\,, \tag{18}$$

which is probably the most informative of all the derived spectra. It will be shown later that, if $y(t)$ is a linear function of $x(t)$, $y(t) = \int h(t-s)\,x(s)\,\mathrm{d}s$, or equivalently, if $x(t)$ and $y(t)$ are both linear functions of another random process, then $\kappa_i = 1$ for every i. On the other hand, if x and y are independent processes then κ_i is always zero. These are two extreme cases. In the first, one process determines the other uniquely; in the second, not at all. An in-between case is

$$y(t) = \int h(t-s)\,x(s)\,\mathrm{d}s + n(t)\,,$$

where $n(t)$ is some random noise completely independent of x. For such two processes x and y, κ_i is close to 1 at those frequencies where y is mainly under the influence of x, and close to 0 where the noise dominates. A more quantitative statement of this relation will be made when we discuss spectrum interpretation.

Additional relations*

This subsection presents a few more equations related to the statistical properties of X. They are not essential for the main stream of our study, but will be required later when studying other topics. The reader may skip it for now and return to it later.

Let us look at the quantity

$$\langle X_i^* X_j \rangle = \int_0^1 \exp(cis)\,\exp(-cjt)\,a(\,|s-t|\,)\,\mathrm{d}t\,\mathrm{d}s\,.$$

By changing the variable of integration to $t' = (t-s)$, dividing the inner integral into two parts, and inverting the order of integration we have

$$\langle X_i^* X_j \rangle = \int_0^1 \exp(-cjt)\,a(t) \int_0^{1-t} \exp[c(i-j)s]\,\mathrm{d}s$$

$$+ \int_{-1}^0 \exp(-cjt)\,a(t) \int_{-t}^1 \exp[c(i-j)s]\,\mathrm{d}s\,. \tag{19}$$

If $i = j$ we recover the expression for S_i given in a. For $i \neq j$ we get

$$\langle X_i^* X_j \rangle = \int_0^1 \exp(-cjt)\,a(t)\,\frac{\exp[c(i-j)(1-t)]-1}{c(i-j)}\,\mathrm{d}t$$

$$+ \int_{-1}^0 \exp(-cjt)\,a(t)\,\frac{\exp[c(i-j)]-\exp[c(i-j)(t-1)]}{c(i-j)}\,\mathrm{d}t$$

* May be omitted on a first reading.

$$= \int_0^1 a(t)[\exp(-cit) - \exp(-cjt)]/c(i-j)\, dt$$

$$+ \int_{-1}^0 a(t)[\exp(-cjt) - \exp(-cit)]/c(i-j)\, dt\,,$$

where we have used $\exp[c(i-j)] = 1$. Now we change the integration variable in the second term to $-t$, and use the fact that $a(-t) = a(t)$, we have

$$\langle X_i^* X_j \rangle = \int_0^1 a(t)[\exp(cjt) - \exp(-cjt) - \exp(cit) + \exp(-cit)]/c(i-j)\, dt$$

$$= \frac{1}{\pi(i-j)} \int_0^1 a(t)[\sin(2\pi jt) - \sin(2\pi it)]\, dt\,,$$

for $(i-j)$ = non-zero integer. (20)

This is usually small, first because $a(t)$ is largest at $t = 0$ but $\sin(0) = 0$, then because of the $1/(i-j)$ term, and finally because of the subtraction of the two sines. Consequently, X_i^* and X_j are approximately uncorrelated in most cases.

We have so far assumed that $(i-j)$ is an integer. Recall, however, that we need both integer and half-integer frequencies to recover $a(t)$ from S. When i and j are both half integers, (20) remains valid, but if only one is, then it does not because $\exp[c(i-j)] = -1$ rather than 1. With this change we get

$$\langle X_i^* X_j \rangle = - \int_0^1 a(t)[\exp(cjt) + \exp(-cjt) + \exp(cit) + \exp(-cit)]/c(i-j)\, dt$$

$$= \frac{\sqrt{-1}}{\pi(i-j)} \int_0^1 a(t)[\cos(2\pi jt) + \cos(2\pi it)]\, dt\,,$$

where $(i-j)$ = odd multiple of ½. (21)

Note that this is purely imaginary, while (20) is purely real. Expression (21) is likely to give larger value than (20) as $\cos(0) = 1$.

Now let us consider $\langle X_i X_j \rangle$. Since we know $X_i^* = X_{-i}$, we obtain $\langle X_i X_j \rangle$ simply by replacing i with $-i$ in (20) and (21). This gives

$$\langle X_i X_j \rangle = \frac{-1}{\pi(i+j)} \int_0^1 a(t)[\sin(2\pi jt) + \sin(2\pi it)]\, dt$$

for $(i-j)$ = non-zero integer, and

$$\langle X_i X_j \rangle = \frac{1}{\pi(i+j)} \int_0^1 a(t)[\cos(2\pi jt) + \cos(2\pi it)]\, dt$$

for $(i-j)$ = odd-multiple of ½. (22)

Equations (22) are true even for $i = j$ (whereas (20) does not as it contains the term $(i-j)$ in the denominator), and gives

$$\langle X_i^2 \rangle = -\frac{1}{\pi i} \int_0^1 a(t) \sin(2\pi it) \, dt. \tag{23}$$

This applies to both integers and half integers. Like (20), this is usually small. By substituting this back into (20) and (22), we get additional relations, such as

$$\langle X_i^* X_j \rangle = [i\langle X_i^2 \rangle - j\langle X_j^2 \rangle]/(i-j), \qquad \text{for } (i-j) = \text{non-zero integer,}$$

and so on. However, these do not seem to have any use for our study and so will not be pursued.

We said earlier that when a window cutoff T is used, we would require the frequencies $i/2T$. This converts (19) into the form

$$\langle X_{i/2T}^* X_{j/2T} \rangle = \int_0^T \exp(-cjt/2T) \, a(t) \, w(t) \int_0^{1-t} \exp[c(i-j)s/2T] \, ds \, dt$$

$$+ \int_{-T}^0 \exp(-cjt/2T) \, a(t) \, w(t) \int_{-t}^1 \exp[c(i-j)s/2T] \, ds \, dt.$$

This leads to more complicated expressions than (20-22) because e $\exp[c(i-j)/2T]$ is not necessarily ± 1. Again we do not pursue this, but the reader might try to carry the derivation further just for fun.

6 Estimation

Estimation of averages

Having introduced in the previous chapter a number of averages and stated that they will be useful, we now consider how we can actually obtain them from measured values of a random process. This is termed *estimation*. That is, we are taking guesses at what the averages should be. We cannot, of course, always guarantee that the guesses will be good. In fact, we said earlier that, even if we observe a random process from the infinite past to the infinite future, we are still not assured of being able to obtain the correct averages unless the process is ergodic. Aside from that, to find good estimates is in all cases a tricky business. It is usual to get some indication of the quality of an estimate by assuming some model for the random process and then computing the average deviation of the estimate from the desired value for that model. But then the goodness of the estimate rests on the goodness of the model assumed. Certain types of estimates are reasonably good for many different kinds of models, and they are said to be *robust*. However, robustness, like beauty, is often in the eyes of the beholder. There is no such thing as a universally robust estimate.

Let us consider how to estimate the mean of a random process, $\mu = \langle x(t) \rangle$. First, let us assume that we have measured $x(t)$ once only. Would this value, say $x(s)$, be a good estimate for μ? Clearly, on average $x(s)$ is μ. But this means very little by itself. All this guarantees is that $x(s) - \mu$ is as likely to be too big as too small. On the other hand, if the mean square deviation is small, then the estimate would be a good one. Thus, let us compute $\langle (x(s) - \mu)^2 \rangle$. By definition this is just the variance of x, $V(x)$, or σ^2. In other words, if x is a random process with small standard deviation, then measuring one value is likely to yield a good estimate of the mean. This result is of course intuitively obvious. However, the foregoing illustrates the procedure for analysing the quality of estimates.

An estimate, say ξ, which, on average, equals the quantity it seeks to estimate, say μ, is said to be *unbiased*. Otherwise it is biased. The difference $\beta = \langle \xi \rangle - \mu$ is called the bias. We write $\langle (\xi - \mu)^2 \rangle$ as $d(\xi)$. This measures the quality of the estimate. As we saw before, the variance of the estimate is as important as the bias for determining the quality of the estimate. To show this more clearly, let us consider the mean square deviation of ξ from μ, but this time assuming that $\langle \xi \rangle \neq \mu$. We now have

$$d(\xi) = \langle [\xi - \langle \xi \rangle - (\mu - \langle \xi \rangle)]^2 \rangle$$

$$= \langle (\xi - \langle \xi \rangle)^2 \rangle + \beta^2 - 2 \langle (\xi - \langle \xi \rangle) \rangle \beta . \tag{1}$$

The first term is just the variance of ξ, while the last term is always 0, so that

$$d(\xi) = V(\xi) + \beta^2 . \tag{2}$$

In other words, the mean square deviation of ξ from μ is the sum of the variance of ξ and the square of its bias. Frequently, one has to trade off between bias and variance in order to get the best estimate. A biased estimate is sometimes superior to an unbiased one, which means that the intuitively accepted estimate is not necessarily the good one. For example, let us examine the possibility of estimating $\mu = \langle x(t) \rangle$ by $\xi_\mu = \alpha x(s)$, $0 \leqslant \alpha \leqslant 1$. We have

$$\beta = \langle \xi_\mu \rangle - \mu = (\alpha - 1)\mu \qquad \text{and} \qquad V(\xi) = \alpha^2 V(x) = \alpha^2 \sigma^2 .$$

Thus

$$d(\xi) = \alpha^2 \sigma^2 + (\alpha - 1)^2 \mu^2 .$$

Fig. 6.1

This is minimized if we take $\alpha = \mu^2/(\sigma^2 + \mu^2)$. $\alpha = 1$ only when $\sigma = 0$. Thus, the biased estimate is better in almost every case. To get a clearer picture of the reason for this, let us suppose that the unbiased estimate falls in the range $\mu - e$ to $\mu + e$. Obviously, the biased estimate would fall in the range $\alpha(\mu - e)$ to $\alpha(\mu + e)$ (Fig. 6.1). However, the overall deviation is now reduced to $2\alpha e$. This occurs because multiplication by α decreases the upper limit more than it decreases the lower limit. We hasten to add that the above theoretical result is useless in practice because we do not know σ, μ to start with, and so cannot find out what α to use.

Now let us suppose we have n measured values $x_1, ..., x_n$. How do we estimate μ and σ from these? It is usual to estimate μ by the *measured mean**

$$\xi_\mu = \frac{1}{n} \sum_{i=1}^{n} x_i , \tag{3}$$

and σ^2 by the measured variance (squared deviation)

$$\xi_{\sigma^2} = \frac{1}{n-1} \sum_{i=1}^{n} (x_i - \xi_\mu)^2 . \tag{4}$$

Both are unbiased estimates. (At first sight it might be thought that we should estimate σ^2 by dividing by n. We shall later see that division by $n-1$ ensures unbiasedness, a fact not intuitively obvious.) It is, however, not at all easy to find their deviation from desired values unless some assumptions about the characteristics of x are made. Thus

$$V(\xi_\mu) = \langle (\xi_\mu - \mu)^2 \rangle = n^{-2} \sum_{i,j} \langle x_i x_j \rangle - \mu^2 ,$$

which depends on the correlation between different measured values. If the measurements have been performed independently of each other then x_i is uncorrelated with x_j if $i \neq j$, which means $\langle x_i x_j \rangle = \langle x_i \rangle \langle x_j \rangle$ so that

$$V(\xi_\mu) = 2n^{-2} \sum_{i} \langle x_i^2 \rangle + n^{-2} \sum_{i \neq j} \langle x_i \rangle \langle x_j \rangle - \mu^2 .$$

Recall that $\langle x^2 \rangle = \mu^2 + \sigma^2$. Thus

$$V(\xi_\mu) = n^{-2} n(\mu^2 + \sigma^2) + n^{-2} n(n-1)\mu^2 - \mu^2 = \sigma^2/n . \tag{5}$$

It is also clear that

$$\langle \xi_\mu^2 \rangle = \mu^2 + \sigma^2/n . \tag{6}$$

Thus, if the values of x are uncorrelated, averaging over n values gives an estimate with mean square deviation of only σ^2/n. (The bias is zero, so there is only the variance term.) This would not be valid, if, say, the measurements had been performed at equally spaced intervals. The requirement of completely random measurements is not easy to satisfy. Consequently, the estimates we apply in real life are seldom as good as indicated by the above analysis.

We shall also examine the statistical properties of ξ_{σ^2}. First we show its unbiasedness, which is not obvious at first sight like that of ξ_μ. We have

$$\langle \xi_{\sigma^2} \rangle = (n-1)^{-1} \sum_{i=1}^{n} \langle x_i^2 + \xi_\mu^2 - 2x_i \xi_\mu \rangle = (n-1)^{-1} (n\langle x^2 \rangle + n\langle \xi_\mu^2 \rangle - 2n\langle \xi_\mu^2 \rangle).$$

Making use of (6) we get

$$\langle \xi_{\sigma^2} \rangle = n(n-1)^{-1} [(\sigma^2 + \mu^2) - (\sigma^2/n + \mu^2)] = \sigma^2 . \tag{7}$$

* The usual name is 'sample mean', but we do not like to have it confused with the sampling method on page 15.

This justifies the division by $(n-1)$. It serves the purpose of compensating for the correlation between x and ξ_μ. Being an unbiased estimate, the mean square deviation of ξ_{σ^2} is just its variance, which is found as

$$\langle(\xi_{\sigma^2} - \sigma^2)^2\rangle = \langle(\xi_{\sigma^2})^2\rangle - \sigma^4 . \tag{8}$$

This expression is extremely difficult to evaluate in general as it involves fourth powers of x. An assumption that simplifies matters greatly is that x by Gaussian, and we shall state without proof that the above expression for Gaussian x is $2\sigma^4/n$. Again by averaging over n uncorrelated measured values we obtain an estimate for σ^2 with reduced mean square deviation. It is also of interest that for both μ and σ^2 there are biased estimates which have smaller mean square deviations than the unbiased estimates. Thus, the best estimate for σ^2 is what one gets by dividing the sum of squares in ξ_{σ^2} by $(n+1)$ rather than $(n-1)$. Of course, when n is large the difference of the two estimates is negligible. In other words, when one is averaging over a large set of random measurements the unbiased estimate tends to approach the optimal estimate. This is because the averaging process produces a smaller variance, and as variance decreases the contribution of the bias towards mean square deviation becomes more important relatively. In such cases the reduction of bias becomes more and more helpful.

The analysis of the statistical properties of estimates computed from correlated measured values is much more difficult. This, unfortunately, is the case when we estimate the spectrum of a random process, as we have to use, for the requirements of both instrumentation and computation, values measured at equally spaced points. It is far from easy to know the quality of estimates, let alone choosing the best. The whole business of choosing a reasonable estimate for the spectrum is very much an art, and a great deal of experimentation is often needed.

In computing an estimate from measured values, the statistical errors of the initial data propagate in very much the same way round-off errors do, ' and there is much similarity between choosing a robust estimate and devising a numerical procedure stable against round-off errors. An estimate involving subtraction of numbers of comparable sizes, for example, is likely to have a large mean square deviation, just as it would also suffer from round off.

Estimation of autocorrelation–statistical properties

In this subsection we consider the procedure of estimating

$$a(t) = \langle x(t+s)\, x(s)\rangle ,$$

from measured values of x. This is an intermediary step before the computation of the spectrum. The autocorrelation contains all the information about a random process that can be extracted from the spectrum, but in most cases the spectrum is easier to interpret and so is to be preferred. However, for some applications, especially those of interest to social scientists and economists, the autocorrelation can be a more direct and informative quantity. We saw its use on page 47 in fitting an autoregressive process to a set of measured values.

To start with, we could use simply $x(t+s)x(s)$ for some particular s as an estimate for $a(t)$, but this in general would not be a good estimate. A better estimate would be the average of $x(t+s)x(s)$ for a set of randomly chosen values of s, $s_1, s_2, ..., s_n$. The difficulty is that the data available can be extremely limited, and to choose a set out of this amounts to throwing away the rest of the precious collection. For better or for worse, we often have to utilize all the data we are given.

The usual thing to do is to take the time average

$$\xi_a(t) = \alpha \int_0^{1-t} x(s)\,x(s+t)\,ds\,, \tag{9}$$

and use that as estimate for $a(t)$. If we are given N equally spaced sample values, then we would use

$$\xi_a(t_i) = \frac{\alpha}{N} \sum_{k=0}^{N-1} x(s_k + t_i)\,x(s_k)\,. \tag{10}$$

The factor α needs some explanation. Obviously, as

$$\langle \xi_a(t) \rangle = \alpha \int_0^{1-t} a(t)\,ds = a(t)(1-t)\alpha\,,$$

a choice of $\alpha = (1-t)^{-1}$ would give us an unbiased estimate. For the sampled value case, $\alpha = N/(N-k)$ yields unbiasedness. However, we shall later see that it is usually better to choose α differently.

Let us study the statistical errors of $\xi_a(t)$. First, it has bias

$$\beta(t) = \langle \xi_a(t) \rangle - a(t) = a(t)[\alpha(1-t) - 1]\,, \tag{11}$$

while its variance is

$$V(\xi_a) = \langle \xi_a^2 \rangle - \langle \xi_a \rangle^2 = \alpha^2 \int_0^{1-t} \langle x(s)\,x(t+s)\,x(s')\,x(t+s') \rangle\,ds\,ds' - \langle \xi_a(t) \rangle^2\,.$$

This expression, again involving fourth-order terms, is difficult to handle for general x. But, assuming x to be Gaussian, we apply equation (22) on page 40, and turn the above into

$$V(\xi_a) = \alpha^2 \int_0^{1-t} [a(t)^2 + a(s-s')^2 + a(t+s'-s)\,a(t+s-s')]\,ds\,ds' - \langle \xi_a(t) \rangle^2\,.$$

The first term in the integrand gives $[\alpha(1-t)\,a(t)]^2$, which cancels the last term. Then

$$V(\xi_a) = \alpha^2 \int_0^{1-t} ds \int_{s+t-1}^{s} a(s'')^2 + a(t+s'')\,a(t-s'')\,ds''\,,$$

where we have changed variable of integration to $s'' = s-s'$. Division of the inner integral into two parts, for $0 \leqslant s'' \leqslant s$ and $s+t-1 \leqslant s'' \leqslant 0$, and then inverting the order of integration gives

$$V(\xi_a) = \alpha^2 \int_{t-1}^{1-t} (1-t-|s''|)[a(s'')^2 +a(t+s'')\,a(t-s'')]\,ds'' . \qquad (12)$$

(The procedure is analogous to that used on page 51 to derive the relation between S and a.) Combining the bias and the variance, we get the mean square deviation ξ_a from $a(t)$ as

$$d(\xi_a) = V(\xi_a) +\beta(\xi_a)^2 = \alpha^2 \int_{t-1}^{1-t} [a(s)^2 +a(t-s)\,a(t+s)](1-t-|s|)\,ds$$

$$+a(t)^2\,[\alpha(1-t)-1]^2 . \qquad (13)$$

If we choose $\alpha = (1-t)^{-1}$, the unbiased estimate, the second term disappears, but the variance term becomes larger, especially for those $a(t)$ with t close to 1. Unfortunately, for most problems the variance term is the more important one, as it includes values of $a(t)$ for all t, including $t \sim 0$, where $a(t)$ tends to be largest. This is why α should be chosen differently. In particular, it should not be allowed to increase with t as fast as $1/(1-t)$. However, it is rather hard to tell what the best α would be because the integral depends on the exact shape of $a(t)$. Usually, it is advisable to take $\alpha = 1$. This is not only the simplest. It also has the advantage that the relation between ξ and Ξ then is just a simple Fourier transformation (page 53). In any case, the decision turns out to be relatively unimportant for computing the spectrum, because we would modify the autocorrelation by a window function, which has a much larger effect on the computed spectrum than a different choice of α. However, if one required the autocorrelation as much as the spectrum, then the choice of α is important. It has been noted on many occasions that the unbiased estimate $a(t)$ is more oscillatory than it should be, since taking $\alpha = 1/(1-t)$ amounts to dividing by a decreasing number as t approaches 1. The oscillations damp out much better if α is chosen to be 1.

Regardless of what value one chooses for α, $\xi_a(t)$ for t near 1 is likely to deviate significantly from $a(t)$. Fig. 6.2 compares $a(t)$ for a second-order auto-regressive process with $\xi_a(t)$ computed by generating a set of values in accordance with its probability distribution and then making use of equation (10), with α chosen to be 1. It can be seen from the figure that the estimate tends to be good around $t = 0$ and becomes progressively worse as t increases. This means that spectral estimates computed from the lower end of the autocorrelation only would be more reliable than one computed using all the values. This can be achieved by choosing a window with a short cutoff. We shall see this practice in the next chapter. .

Estimation of autocorrelation: procedures

We have seen that the simplest estimate for $a(t)$ is the so-called lagged product sum with $\alpha = 1$

$$\xi_a(t_i) = \frac{1}{N} \sum_{j=0}^{N-i-1} x(t_k)\, x(t_k + t_i) , \tag{14}$$

and that it is more reliable for small i than large i. Quite often it happens that, although we have N sampled values of $x(t)$ we do not bother to estimate all N values of the autocorrelation, since we would be using a window with cutoff $\pm T$, so that i does not exceed some value $(M-1)$, $M = TN$. There are thus M values of $\xi_a(t)$ which we use to compute the spectrum. Let us now consider the procedures for actually evaluating these M values.

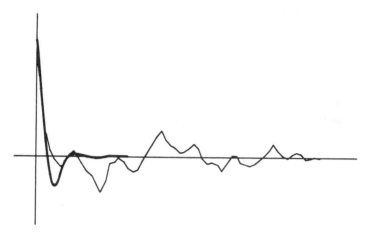

Fig. 6.2

Direct method. To compute $\xi_a(t_i)$ directly from (14) takes $(N-i)$ multiplications and additions, so that evaluation of M of these requires $\sum_{i=0}^{M-1} (n-i) \sim \epsilon MN$ operations, $\epsilon \sim \frac{1}{2}$ if $M \sim N$, $\epsilon \sim 1$ if $M \ll N$. When N and M are both large, this is rather laborious. The alternative, the FFT method, is supposed to require only $N(\log M)$ operations. This has led many people to believe that the direct method is now obsolete. We disagree. In fact, the direct method is clearly superior if $M < 256$, regardless of what N is. Even for M approaching 1000 the direct method may still be preferred.

An important advantage of the method is its simplicity and its robustness against round-off errors. The first is obvious. The second advantage comes about because in (14) we add up $(N-i)$ products. This causes cancellation between the round-off errors of the products, such that N round-off errors added up would produce an overall error about $N^{\frac{1}{2}}$ times that of each product. Since the sum is to be divided by N, the final round-off error is only $N^{-\frac{1}{2}}$ times the initial error. If, say, the x values are fixed point numbers of m bits, then the final round-off error is about $\sigma_x^2\, 2^{-m-n/2}$, where $n = \log_2(N-i)$.

As a result of this robustness, we can choose m to be fairly small but still end up with reasonably accurate results. While digitized signals usually have between 8 to 12 bits, for the purpose of correlation estimation 6 bits or even less are already quite adequate in most cases. (In fact, if the signal is Gaussian one can recover the autocorrelation quite accurately even when each number is 'clipped', i.e., reduced to *one bit* each, indicating whether the number is + or −. Of course, we do not recommend doing this in practice unless one is absolutely sure that the signal is Gaussian, which is not always exactly true.) This reduces the data storage needs considerably, and can also reduce computing time since now the product of two x values can be stores in a single minicomputer word (usually 16 bits) instead of two, and the sum accumulation is accordingly speeded up. Of course, multiplying two 6-bit numbers is in theory much faster than multiplying two 10-bit numbers, but it is rather hard to take advantage of this without employing special hardware.

A large core requirement is a supposed disadvantage of the direct method, since, it is claimed, computation of each $\xi_a(t_i)$ requires all x_k. Actually, we do not need any more than $2M$ core locations, M for the ξ values and M for the x. We shall call the first M locations A_k, $k = 0, 1, ..., M-1$, and the second M, B_k. The actual procedure goes as follows: Clear the words A_k for every k, and read $x_0 ... x_{M-1}$ into B_k. Compute the products $x_0 x_0$, $x_0 x_1, ..., x_0 x_{M-1}$ and add them to $A_0 ... A_{M-1}$. Because x_0 is now no longer needed, x_M is read into B_0, overwriting x_0. The products $x_1 x_1$, $x_1 x_2, ..., x_1 x_M$ are then computed and added to $A_0 ... A_{M-1}$. This frees the address containing x_1, into which we read x_{M+1}. This then allows all the products involving x_2 to be computed and added to A_k. And so on.

FFT method. We saw on page 9 that multiplication of the Fourier transforms of two numbers produces a cyclic convolution. If we multiply the complex conjugate of a Fourier transform with another transform we produce correlation. That is

$$c(t) = \sum_i \exp(cit)X_i Y_i^* = \sum_i X_i \exp(cit) \sum_j Y_j^* \int \exp[c(i-j)s] \, ds,$$

the last integral being just δ_{ij}. We then have

$$c(t) = \int \sum \exp[ci(t+s)] X_i \sum Y_j^* \exp(-cjs) \, ds$$

$$= \int x(t+s)[\sum Y_j \exp(cjs)]^* ds = \int x(t+s) \, y(s) \, ds,$$

again remembering that if $(t+s)$ goes outside the range $[0, 1]$ x is assumed to be periodically extended. The above expression is called a cyclic correlation. More explicitly, this is

$$c(t) = \int_0^{1-t} x(t+s) \, y(s) \, ds + \int_{1-t}^1 x(t+s-1) \, y(s) \, ds. \qquad (15)$$

Suppose $y(s) = 0$ for $\tau \leqslant s \leqslant 1$, then the second term vanishes if $t < 1-\tau$, and $c(t)$ reduces to the ordinary correlation between x and y. In other words, we can compute the lower end of the correlation by multiplication of Fourier transforms if the higher end of $y(t)$ is zero. Even when the given functions do not have identically zero sections, we can still use this technique simply by adding enough zeros to the two functions.

Thus, given N-element vectors x and y, to compute the first M values of their correlation we simply add M zeros to each vector, compute their $(N+M)$-point FFT, multiply these, and take the inverse transform of the products. The first M values of the inverse transform give the M required correlations. (The remaining N numbers are of no use, and are simply thrown away.) If we want all N values of the correlation, then it would be necessary to add N zeros to each of x and y, doubling the vector size.

In theory, the FFT method requires $(N+M)\log_2(N+M)$ operations, and so is faster than the lagged product method. In practice, it does not work out so well. To start with, two FFT's are needed. Second, $(N+M)$ is not always a power of 2. To be able to use FFT, we must add more zeros to the data to bring the vector size of 2^n. In the lagged product method, we simply multiply and add real numbers; FFT requires complex arithmetic, almost certainly in floating point. Further, taking 2^n-point FFT takes at least 2^n core locations for data storage, though there is a way of overcoming this as shown below. Finally, the process is not as robust against round-off errors. Finding accurate correlations takes input data of comparable precision. Six-bit quantization will not work.

For these reasons we feel there is little advantage in the FFT method when M is relatively small. Taking $N = M = 256$, the lagged product method takes $N^2 = 2^{15}$ operations to compute all the correlations, while the FFT method takes, in theory, about $2.9.2^9 \simeq 2^{13}$ operations. But when one takes into account program simplicity, real versus complex arithmetic, etc., the supposed advantage of the FFT method evaporates. Taking a larger N affects the comparison only marginally. However, if M becomes larger than the FFT method would be favoured.*

When N is very large, it may become impossible to accommodate all the elements of x or y in core to carry out the FFT computation. The solution is to divide the data into shorter blocks and compute the correlation between these, and then add up the results. The procedure for correlating x and y is as follows:

1. Divide x and y into N/M blocks of M elements each. Label these as $x^1, x^2, ..., x^{N/M}$ and $y^1, y^2, ..., y^{N/M}$, with

2. Form $2M$-element vectors u^i and v^i, $i = 1, 2, ..., N/M$, with

$$v_k^i = x_k^i, \quad u_k^i = y_k^i, \quad k = 0, 1, ..., M-1 ,$$

$$v_k^i = x_k^i, \quad u_k^i = 0 , \quad k = M, M+1, ..., 2M-1 .$$

In other words, each block of u contains one block of y and one block of zeros, while each block of v has two blocks of x.

* The remainder of the section may be omitted.

3. Compute the FFT of v^i and u^i. Call these V^i and U^i. Multiply V^i and U^{i*} element by element and take the inverse transform. Each block of u, v gives an inverse transform of $2M$ values. Keep only the first half of each inverse transform, calling this c^i. (This should be *real*.)

4. Compute $\xi_k = \sum_{i=1}^{N/M} c_k^i$ to give the kth term of the required correlation.

When we are taking autocorrelation, y and x are the same vector. The procedure becomes somewhat simplified:

1. Divide x into N/M blocks $x^1, ..., x^{N/M}$.

2. Form $2M$ elements u^i as before.

3. Compute U^i, the FFT of u^i. Form V^i as follows: $V_k^i = U_k^i + (-1)^k U_k^{i+1}$. Multiply V^i and U^{i*} and take the inverse transform. Keep the first half c^i.

4. $\xi_k = \sum_{i=1}^{N/M} c_k^i, \ k = 0, 1, ..., M-1.$

The main difference is of course that now $x = y$ so that v^i is just u^i and u^{i+1} combined, which means that V^i can be found by combining U^i and U^{i+1}.

The method outlined reduces storage requirement at a slight increase in computing time. Again, this is not recommended unless M is well over 256.

Statistical properties of spectral estimates

We have defined S_i as $\langle |X_i|^2 \rangle$ so it would be natural to estimate it simply by takine $|X_i|^2$. This is the *periodogram* of the given function $x(t)$. Let us study its statistical properties. We shall again denote an estimate for S_i as Ξ_i, as in Chapter 5. We saw

$$\Xi_i = \int_0^1 \exp[ci(t-s)] \, x(t) \, x(s) \, dt \, ds \, .$$

Obviously, by definition, Ξ_i is an unbiased estimate for S_i so that its mean square deviation from S_i is just its variance, which is found as follows:

$$V(\Xi_i) = \langle \Xi_i^2 \rangle - S_i^2 = \int \exp[ci(t-s) + ci(t'-s')] \langle x(t) x(s) x(t') x(s') \rangle \, dt \, ds \, dt' ds' - S_i^2.$$

Again we make the simplifying assumption that x is Gaussian, so that the use of (22) in Chapter 4 yields

$$\langle x(t) x(s) x(t') x(s') \rangle = a(t-s) \, a(t'-s') + a(t'-s) \, a(t-s') + a(t-t') \, a(s-s') \, .$$

The first two terms integrated with $\exp[cit(t-s) + ci(t'-s')]$ gives $2S_i^2$, so that

$$V(\Xi_i) = 2S_i^2 + \int \exp[ci(t+t')] \langle x(t) x(t') \rangle \, dt \, dt' \int \exp[-ci(s+s')]$$
$$\langle x(s) x(s') \rangle \, ds \, ds' - S_i^2$$

$$= S_i^2 + |\langle X_i^2 \rangle|^2 \, .$$

When x is white noise, which implies $\langle x(t)\,x(t')\rangle = \delta(t-t')$, the second term vanishes. In most other cases the second term is quite small (see page 62). This means that $V(\Xi_i)$ is approximately S_i^2, or, Ξ_i is likely to deviate from S_i by about S_i! Clearly, Ξ_i is most unlikely to be a good estimate for S_i. Fig. 6.3 compares S with Ξ computed from the same values as those used in Fig. 6.2.

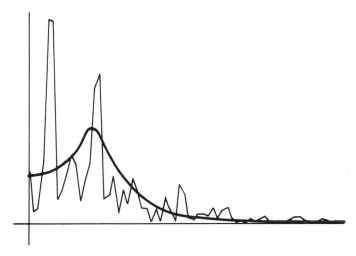

Fig. 6.3

The reason for this problem is not hard to see. We saw on page 53 that

$$\Xi_i = \int_{-1}^{1} \exp(-cit)\,\xi_a(t)\,\mathrm{d}t.$$

Earlier, we saw that the quality of $\xi_a(t)$ becomes progressively worse as t approaches 1. The inclusion of such values in Ξ_i adversely affects its quality. Consequently, to obtain a better estimate for S_i one must reduce the dependence of Ξ_i on $\xi_a(t)$ with t close to 1. This is again achieved by appropriate choice of the window function, which serves two purposes, leakage suppression and reduction of statistical instability.

The use of a window gives spectral estimates of the form of (14) in Chapter 5, Ξ^w, which is the Fourier coefficient of the product $w(t)\,\xi_a(t)$. Previous results (page 8) state that multiplication in time space corresponds to discrete convolution in frequency space. If we write the Fourier transform of $w(t)$ over $[-1, 1]$ as $\overline{W}$, then we have

$$\Xi_{i/2}^w = \sum_k \overline{W}_{\frac{1}{2}k}\,\Xi_{\frac{1}{2}(i-k)}. \tag{17}$$

In other words, Ξ^w is Ξ averaged from i minus some value to i plus some value, with appropriate weight for each Ξ. There is, of course, no reason why we have to obtain Ξ^w indirectly by first computing ξ_a. It is equally feasible to obtain Ξ^w from Ξ according to (17), provided $\overline{W}$ is non-zero for a small number of k only so that only a few additions and multiplications are required for each Ξ_i^w. This is the moving average method for obtaining stable spectral estimates, (17) being a moving average on Ξ. It is also called the smoothed periodogram method, as taking moving averages smoothes out the fluctuations in Ξ.

Equation (17) is a convolution of Fourier coefficients defined over $[-1, 1]$. Now, if the $w(t)$ is non-zero for $-T \leqslant t \leqslant T$ only, then we can also define Fourier coefficients over $[-T, T]$. However, these coefficients correspond to frequencies $i/2T$. Or, we have

$$\Xi_{i/2T}^w = \sum_k \overline{W}_{k/2T} \, \Xi_{(i-k)/2T} . \tag{18}$$

In other words, we can also evaluate Ξ^w by first Fourier transforming $\xi_a(t)$ over $[-T, T]$ to produce $\Xi_{i/2T}$, and then using (18). However, this would be convenient only if both $w(t)$ and $\overline{W}_{k/2T}$ are non-zero for only a small set of values. As we shall see, this condition is true for the Tukey windows.

Another approach is that of segment averaging. The idea is to divide up the interval over which $x(t)$ is measured into $1/T$ segments. A discrete Fourier transform is computed from the values of x in each segment. This then produces a set of periodograms, which are of course unstable. The average of these periodograms, however, is more stable, and is usable as spectral estimate. We shall study this method later. It is special in that Ξ^w computed this way is not related to Ξ by (18).

Already we have seen several different ways for finding Ξ^w: multiply $w(t)$ into $\xi_a(t)$ and transform; truncate ξ_a between $[-T, T]$, transform and then use (18); compute the periodogram and use (17); segment. Each needs a particular windowing method.

Conceptually, the moving average method is by far the easiest to understand. The method appears so simple that it often gives the reader a sense of acute disappointment – why should anyone study random processes, estimation errors, etc., only to learn to use *that*? The reason is that the method needs some finesse to apply if one is interested in the best results for the lowest cost. A little bit of theory does the reader no harm.

7 Windowing

Time and frequency windows

We saw in Chapter 2 that the Fourier transform of a sinusoid with a non-integer frequency is not a Kronecker delta because the function has been abruptly terminated, giving the so-called leakage effect. Windowing is essentially the gradual termination of a function in time space. By appropriately choosing this time window we can obtain two benefits. First, leakage can be confined to frequencies close to the true frequency. This result depends mainly on the shape of the window, and the art of choosing a good shape is called *window carpentry*. Second, it is also possible to average the power of the Fourier components over some range of frequencies. This reduces the resolution of the computed spectrum, since we can no longer hope to see the individual components, but is essential for producing a reliable spectrum. This result depends on the length of the window, and the art of choosing a suitable length is called *window closing*.

It may seem strange that we should try to see the true shape of a Fourier transform by altering its time function with a window. We can give two reasons as to why this in fact does work. First, even when we simply Fourier transform a function without alteration, a window is still implicitly present, and it happens to be a bad window. The window would reproduce a transform with only integer frequency components exactly, but does badly for a more general transform. By choosing another window, we hope to do better in general. Second, the Fourier transform of a random process contains many statistical fluctuations, which considerably exceed the distortions introduced by a window. By smoothing out these fluctuations, the window does much more good than harm.

Before the FFT became widely known spectrum was computed from the autocorrelation estimate $\xi_a(t)$, and windowing was performed by multiplying this by $w(t)$ as we mentioned several times in the last two chapters. With the autocorrelation method now somewhat out of fashion the old window-ing methods became unpopular too. Not only did many people feel the need to devise new windows, it was also suggested that one should perform windowing in both time and frequency spaces. In time space, $x(t)$, not $\xi_a(t)$, is modified by the multiplication of $w(t)$ for the purpose of leakage correction, while weighted averaging on the spectrum is used to reduce variance. We believe, however, that the necessary reform is much less radical. This will be explained as we go along.

We saw on page 13 that the Fourier coefficients of $x(t) = \exp(cft)$ is

$$X_i = \exp(\tfrac{1}{2}cf) \exp(-ci/2) \operatorname{sinc}(i-f), \tag{1}$$

where we have written f for $(j+a)$. The leading term, independent of i, has no effect on the relative importance of X_i. Now let us suppose that $x(t)$ is a linear superposition of a set of sinusoids with unrestricted frequencies:

$$x(t) = \int_{-\infty}^{\infty} X(f) \exp(cft) \, df.$$

We have an integral rather than a sum because f is a continuous variable and is not restricted to any particular range of frequencies, hence the integration limits $\pm\infty$. In reality, of course, $X(f)$ would be non-zero for only some f. It is the 'true' Fourier transform of $x(t)$, but we cannot see it because of our finite time window, which ranges from 0 to 1 at most. Note that $X(f)$ is not unique. That is, two different 'true' Fourier transforms can appear identical through our finite window.

The Fourier coefficients we actually see would be

$$X_i = \int_{-\infty}^{\infty} X(f) \int_0^1 \exp(-cit) \exp(cft) \, dt \, df$$

$$= \exp(-\tfrac{1}{2}ci) \int_{-\infty}^{\infty} \operatorname{sinc}(i-f) \exp(\tfrac{1}{2}cf) X(f) \, df. \tag{2}$$

This is a convolution defined over all possible frequencies. The factor in front, $\exp(-\tfrac{1}{2}ci) = \cos(\pi i) + \sqrt{-1}\sin(\pi i)$, is -1 for odd i and $+1$ for even i, so that it is really $(-1)^i$. Equation (2) shows that taking Fourier series of any function over $[0, 1]$ corresponds to viewing the true spectrum $X(f)$ through the spectral window $(-1)^i \operatorname{sinc}(i-f) \exp(\tfrac{1}{2}cf)$.

Now let us take instead the windowed Fourier coefficients

$$X_i^w = \int_0^1 \exp(-cit) \, w(t) \, x(t) \, dt. \tag{3}$$

Let us write the ith Fourier coefficient of $w(t)$ as W_i. We know that X^w is the discrete convolution between X and W, which gives

$$X_i^w = \sum_j X_{i-j} W_j = \sum_j W_j \int_{-\infty}^{\infty} X(f) \exp(\tfrac{1}{2}cf)(-1)^{i-j} \operatorname{sinc}(i-j-f) \, df,$$

where we have used the expression (2) for X_{i-j}. This can be written as

$$X_i^w = \int_{-\infty}^{\infty} X(f)\, W(i-f)\, df, \tag{4}$$

with

$$W(i-f) = \exp(\tfrac{1}{2}cf) \sum_j W_j(-1)^{i-j} \operatorname{sinc}(i-f-j). \tag{5}$$

In other words, by computing X_i^w we are now viewing $X(f)$ through a different spectral window.

Imagine that our $W(i-f)$ is $\delta(i-f)$. Integrating this with $X(f)$ would produce $X_i^w = X(i)$, so that $X(f)$ is now faithfully reproduced. Unfortunately, this perfect ideal cannot be achieved. The best we can ask for is that $W(i-f)$ have a fairly 'concentrated' shape, with no appreciable oscillations, so that it would produce a somewhat 'smeared out' but still good image of $X(f)$. The absence of oscillations ensures that a sharp peak in $X(f)$ would not produce a series of oscillating peaks in X^w. Our task in window carpentry is then to choose $w(t)$, or equivalently, to choose its Fourier coefficients W_j, such that $W(i-f)$ has a good shape. Expression (5) specifies $W(i-f)$ in terms of W_j. We can also obtain the former directly from $w(t)$. Equation (1) shows that $(-1)^{i-j} \operatorname{sinc}(i-j-f)$ is just the $(i-j)$th Fourier coefficient of $\exp(cft)$. It follows that

$$W(i-f) = \int_0^1 \sum_j W_j \exp[-c(i-j)t]\, \exp(cft)\, dt = \int_0^1 w(t)\exp[-c(i-f)t]\, dt. \tag{6}$$

Thus, W_j is really just $W(i-f)$ at an integer argument $(i-f) = j$. We might consider the former to be 'sampled values' of the latter. Equation (5) can be looked upon as a sinusoidal interpolation in terms of sampled values like (33) on page 16.

Since $X_i^w = \sum_j X_{i-j}\, W_j$, a discrete convolution, we could compute X^w after Fourier transforming $x(t)$, instead of multiplying by $w(t)$ before Fourier transformation. The effect of either is a continuous convolution (3). In summary, multiplication or discrete convolution is how we carry out the computation; continuous convolution is what we achieve in effect.

Now let us look at windowing in computing the spectrum. One way of performing it is to define

$$\Xi_{i/2T}^w = \int_{-T}^{T} \exp(-cit/2T)\, w(t)\, \xi_a(t)\, dt/2T.$$

As explained on page 53, $w(t)$ vanishes outside $[-T, T]$ so that it is sufficient to compute terms for frequencies $i/2T$. We denote the Fourier coefficients of $w(t)$ defined over $[-T, T]$ by $\overline{W}$, and we have

$$\Xi_{i/2T}^w = \sum_j \Xi_{(i-j)/2T}\, \overline{W}_{j/2T}. \tag{7}$$

This gives us the second method for performing windowing.

Suppose that $\xi_a(t)$ can be represented in terms of its 'true' transform:

$$\xi_a(t) = \int\limits_{-\infty}^{\infty} \exp(cft)\, \Xi(f)\, df,$$

then we also have

$$\Xi_{i/2T}^w = \int \overline{W}(i/2T-f)\, \Xi(f)\, df, \tag{8}$$

analogous to (4), with

$$\overline{W}(i/2T-f) = (2T)^{-1} \int\limits_{-T}^{T} \exp[-c(i/2T-f)t]\, w(t)\, dt. \tag{9}$$

However, the relation between $\overline{W}(i/2T-f)$ and $\overline{W}_{j/2T}$ is much simpler than (5). We change the variable of integration to $t' = t/2T$. This converts the limits of integration to $\pm\frac{1}{2}$. Thus, expressing $w(t)$ by its Fourier series over $[-T, T]$ we get

$$\overline{W}(i/2T-f) = \int\limits_{-\frac{1}{2}}^{\frac{1}{2}} \sum_j \overline{W}_{j/2T} \exp(cjt') \exp[-c(i-2Tf)t']\, dt'$$

$$= \sum_j \overline{W}_{j/2T} \frac{\exp[\frac{1}{2}c(i-j-2Tf)] - \exp[-\frac{1}{2}c(i-j-2Tf)]}{c(i-j-2Tf)}$$

$$= \sum_j \overline{W}_{j/2T} \sin[\pi(i-j-2Tf)]/[\pi(i-j-2Tf)]$$

$$= \sum_j \overline{W}_{j/2T} \operatorname{sinc}(i-j-2Tf). \tag{10}$$

Once again, $\overline{W}(i/2T-f)$ is the window through which we see the true spectrum contained in our data, $\Xi(f)$, and we would like it to have concentrated shape and be free of oscillations. We can choose our window by either specifying $w(t)$, which determines $\overline{W}(i/2T-f)$ through (9), or specifying $\overline{W}_{j/2T}$, which determines $\overline{W}$ through (10). There is another way of specifying a window – in terms of $\overline{W}_{j/2}$, the Four transform of $w(t)$ over $[-1, 1]$. This corresponds to putting $T = 1$ in (7), and we compute the windowed spectrum by

$$\Xi_{i/2}^w = \sum_j \overline{W}_{j/2}\, \Xi_{(i-j)/2}. \tag{7a}$$

In other words, such windows are applied by taking averaged values of the *periodogram*, whereas in (7) the window operates on the Fourier transform of $\xi_a(t)$ over $[-T, T]$. Windows specified in this way do not sharply cut off $\xi_a(t)$ for larger values of t, which is why we *could* obtain Ξ^w at half-integer frequencies if we so wish. In actual practice, we do not. Instead, we take i at some widely spaced interval, which is essentially equivalent to using frequencies $i/2T$ for some T. In theory, however, these windows have cutoff ± 1. In any case, the window on the true spectrum is obtained from (10) by putting in $T = 1$, so that we get

$$\overline{W}(i/2-f) = \sum_j \overline{W}_{j/2}\, \operatorname{sinc}(i-j-2f). \tag{10a}$$

Some useful windows

Let us summarize the last section before proceeding: there are three ways of specifying a window. First we could choose an $w(t)$. Or, we could specify its Fourier transform over $[-T, T]$. The former is multiplied into $\xi_a(t)$ *before* Fourier transformation to produce the computed spectrum; the latter is applied as in (7), on the Fourier transform. The third method is to specify the Fourier transform of the window function over $[-1, 1]$. These values operate on the periodogram. Some windows are easier to specify one way; others the other way. In all cases, the purpose is to get a good window $\overline{W}(i/2T-f)$ that operates on the true spectrum $\Xi(f)$. The three types of chosen quantities, $w(t)$, $\overline{W}_{j/2T}$ and $\overline{W}_{j/2}$, determine $\overline{W}(i/2T-f)$ via the equations (9), (10) and (10a).

1. *Rectangular window.* This is specified by $w(t) = 1$, $-T < t < T$; $w(t) = 0$ elsewhere. We use this window if we simply truncate $\xi_a(t)$. (That is to say, even no window is a window.) It gives

$$\overline{W}(i/2T-f) = \int_{-T}^{T} \exp[-c(i/2T-f)t]\ dt = 2T\ \text{sinc}\,(i-2Tf)\ . \tag{11}$$

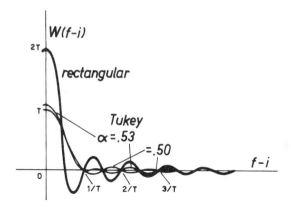

Fig. 7.1

This spectral window is plotted as a function of f in Fig. 7.1. It consists of a series of peaks of alternating signs. The main peak or *main lobe* extends from $f = i - \frac{1}{2}T$ to $f = i + \frac{1}{2}T$, and has height $2T$. The peak on either side has height of around $0.4T$, or about 20% of the main lobe. These and peaks further away are called *side lobes*. Side lobes of this window decrease like $(i/2T-f)^{-1}$ as f moves away from $i/2T$.

Thus, if we compute a spectrum simply by cutting off the bad parts of $\xi_a(t)$, we obtain results related to the true spectrum by

$$\Xi_{i/2T}^{w} = \int_{-T}^{T} \xi_a(t) \exp(-cit/2T)\ dt/2T = \int_{-\infty}^{\infty} \text{sinc}(i-2Tf)\ \Xi(f)\ df\ . \tag{12}$$

Fig. 7.1 shows that these estimates are likely to be poor. With the large, slowly converging and oscillating side lobes, $\Xi^w_{i/2T}$ would include contribution from parts of $\Xi(f)$ very far away from $\Xi(i/2T)$, or, to put it another way, there is much leakage into the frequency interval we are looking at, making the estimates unreliable. It is desirable that $\Xi^w_{i/2T}$ should include only $\Xi(f)$ with f close to $i/2T$, and to this end we would look for windows with more concentrated shapes than the rectangular window.

2. *Bartlett (triangular) window.* From $w(t) = 1 - |t|/T, -T < t < T,$ we have

$$\overline{W}(i/2T - f) = T[\text{sinc}(i - Tf)]^2. \tag{13}$$

This window, shown in Fig. 7.2, is of much historical interest. It is virtually the only window that can be implemented in 'hardware' as done in optics. Its reversed version is none other than the Fejer series discussed on page 12, where we multiply the Fourier coefficients of a function by $1 - |i|/(M+1)$ in order to eliminate the oscillating errors of Gibbs phenomenon. It is also one of the earlier windows used in spectral analysis. However, Bartlett windows have by now gone out of use. Its side lobes, decreasing as $(i - Tf)^{-2}$, do not converge as fast as those of some other windows.

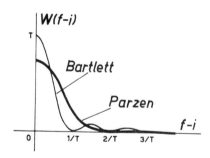

Fig. 7.2

It is interesting to derive (13) as follows: we can obtain $1 - |t|/T$ by convolving the rectangular window of width $[-\tfrac{1}{2}T, \tfrac{1}{2}T]$ with itself and then dividing the result by T. Since convolution in time domain corresponds to multiplication in frequency domain, the Fourier transform of $1 - |t|/T$ is just the square of the transform of the rectangular window, which is $T\text{sinc}(i - Tf)$ if we take into account its shrinked width. Squaring and dividing by T gives (13).

3. *Parzen window.* Here $w(t) = 1 - 6t^2/T^2 + 6|t|^3/T^3$, for $t < T/2$; $w(t) = 2(1 - |t|/T)^3$, for $T/2 < |t| < T$; $w(t) = 0$ elsewhere. Hence

$$\overline{W}(i/2T - f) = \tfrac{3}{4}T[\text{sinc}(i - \tfrac{1}{2}Tf)]^4. \tag{14}$$

We get the Parzen window by convolving $1-2|t|/T$ with itself and then multiplying the result by $\frac{3}{4}T$. Its spectral window is therefore the square of the Bartlett window multiplied by $\frac{3}{4}T$ and stretched along the f axis to twice its original width. The window is also shown in Fig. 7.2.

4. *Tukey window.* Here $w(t) = \alpha +(1-\alpha)\cos(\pi t/T)$, for $-T < t < T$. Hence

$$\overline{W}(i/2T-f) = 2\alpha T \operatorname{sinc}(i-2Tf) +(1-\alpha)\,T[\operatorname{sinc}(i-2Tf+1) +\operatorname{sinc}(i-2Tf+1)] \;.(15)$$

We choose α to be approximately 0.5. If it is exactly 0.5 the window has the much simplified form:

$$\overline{W}(i/2T-f) = T\operatorname{sinc}(i-2Tf)/[1-(i-2Tf)^2] \;. \tag{16}$$

A Tukey window with this choice of α is attributed by Blackman and Tukey to Julius von Hann, calling its application 'hanning'. Using the Tukey window with $\alpha = 0.54$ is called 'hamming' in that paper, in honour of Richard Hamming. This is supposed to be the optimal choice, giving the smallest side lobes. We show in Fig. 7.1 these two Tukey windows along with the rectangular window. Although at first sight the Hamming version appears much more attractive, the advantage is not decisive as the sidelobes converge more slowly as f goes away from $i/2T$. Both windows have a main lobe extending from $(i-2)/2T$ to $(i+2)/2T$, rather than $(i-1)/2T$ to $(i+1)/2T$ as in the case of the rectangular window. The sidelobes oscillate, but they are small and converge rapidly.

Tukey windows are special in that they have simple Fourier transform over $[-T, T]$, with

$$w(t) = \alpha-\frac{1}{2}(1-\alpha)[\exp(ct/2T) +\exp(-ct/2T)] \;,$$

so that $\overline{W}_{j/2T} = \alpha, j = 0; \;\; \frac{1}{2}(1-\alpha), j = \pm 1; \;\; = 0$ otherwise.

Consequently, they are equally simple to apply in time space by multiplication of $w(t)$, or in frequency space by (7), which has the form

$$\Xi^w_{i/2T} = \frac{1}{2}(1-\alpha)[\Xi_{(i-1)/2T} +\Xi_{(i+1)/2T}] +\alpha\Xi_{i/2T} \;.$$

This also explains why (15) is the sum of three sinc functions.

A comparison between the windows of Bartlett, Tukey and Parzen for equal T is of interest here. The side lobes of the Bartlett window are rather large to start with, and they do not converge very fast. However, it has the most concentrated main lobe. The Parzen window has hardly any sidelobes, but its main lobe is rather wide. The Tukey windows fall somewhere in between, and so form a good compromise. They also have the advantage of being easier to apply in frequency space. There is the problem that spectral estimates computed using Tukey windows are not always positive. This is because they have negative side lobes, so that a sharp peak in the spectrum would leak negative power to frequencies near by. While unpleasant, this is not of great consequence, since the spectrum goes negative only if its true value is very small so as to be masked by the negative leakage. Thus, we could simply put negative values to zero. Occasionally, the side lobes 'resonate' with fluctuations in the spectrum, i.e., the spectrum goes up and down at a spacing of abound $1/2T$, and contribution from different side lobes accumulate to produce a false negative peak. But this occurs very rarely. In any case, Tukey windows are the most popular windows.

5. *Cosine taper.* In this case $w(t) = 1$, for $-\frac{4}{5}T < t < \frac{4}{5}T$; $w(t) = \frac{1}{2} + \frac{1}{2}\cos(5\pi t/T)$ for $-T < t < \frac{4}{5}T$ and $\frac{4}{5}T < t < T$; $w(t) = 0$ otherwise. Unlike the other windows, this one does not drop to zero gradually from $t = 0$ on. It is constant for 80% of its duration, and drops off in the 10% at either end. The corresponding spectral window is derived as follows: $w(t)$ is the sum of $w_1(t)$ and $w_2(t)$, with

$$w_1(t) = \frac{1}{2} + \frac{1}{2}\cos(5\pi t/T), \quad \text{for } -T < t < T,$$

and

$$w_2(t) = \frac{1}{2} - \frac{1}{2}\cos(5\pi t/T), \quad \text{for } -\frac{4}{5}T < t < \frac{4}{5}T.$$

Both have simple $\overline{W}_{j/2T}$. The spectral window is approximately

$$\overline{W}(i/2T - f) \sim T \operatorname{sinc}(i - 2Tf) + \frac{4}{5}T \operatorname{sinc}(i - \frac{8}{5}Tf).$$

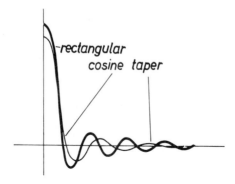

Fig. 7.3

We show the window in Fig. 7.3 together with the rectangular window. Compared with the others this window does not look very attractive. In fact, it is hardly an improvement over the rectangular window. It is not used in spectral estimation via autocorrelation. Why we should have interest in it at all will be explained later.

6. *Daniall window.* This window is specified in terms of $\overline{W}_{j/2}$. It does not have a sharp cutoff T. Its $w(t)$ decreases $\xi_a(t)$ for t near ± 1.

$$\overline{W}_{j/2} = (2m+1)^{-1}, \quad \text{for } j = -m, (-m+1), ..., (m-1), m; \quad = 0 \text{ otherwise}.$$

And it has

$$\overline{W}(i/2 - f) = (2m+1)^{-1} \sum_{j=-m}^{m} \operatorname{sinc}(i - 2f - j).$$

$$(17)$$

This spectral window is shown in Fig. 7.4. It is roughly constant from $f = (i-m)/2$ to $f = (i+m)/2$, and has oscillating side lobes outside this interval. Because of cancellation of contributions from $(2m+1)$ different sinc functions, the side lobes are quite small. Thus, application of this window to the periodogram by (7a) gives results that approximately equal to the average power in a frequency band $(i-m)$ to $(i+m)$, where m plays a part roughly equivalent to $1/T$.

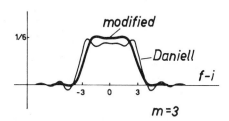

Fig. 7.4

7. *Modified Daniall window.* Here $\overline{W}_{j/2} = 1/2m$, for $(-m+1) \leqslant j \leqslant (m-1)$; $= 1/4m$, for $j = \pm m$; $= 0$ otherwise. This choice of values of $\overline{W}$ provides near maximum cancellation of side lobes, at the cost of making the edges of the window less steep. (See again Fig. 7.4.)

Effects of windowing

We saw on page 73 that the Fourier transform of $\xi_a(t)$, the periodogram Ξ, consists of a set of random variables with large variances. Hence Ξ is virtually useless as an estimation of S. By the use of a suitably chosen window, we can cut off or at least greatly reduce the least reliable portions of $\xi_a(t)$, producing results with much less fluctuation or variance. A different way of looking at it is as follows. Since our spectral estimate is now

$$\Xi_i^w = \sum_j W_j \,\Xi_{i-j} = \int_{-\infty}^{\infty} \Xi(f) \,\overline{W}(i-f) \,\mathrm{d}f,$$

$$(18)$$

we are now averaging over a number of near-independent values. As shown in Chapter 6, this produces results that have smaller fluctuations around their average values, or, putting it differently, the results have less variance or greater statistical stability. However, since the spectral elements being averaged over may have different mean values, we can never hope to exactly reproduce the spectrum. In other words, in general $\langle \Xi^w \rangle \neq S$. Instead,

$$\langle \Xi_i^w \rangle = S_i^w = \int_{-\infty}^{\infty} \overline{W}(i-f) \, S(f) \, \mathrm{d}f.$$

Thus, we say that Ξ^w is a biased estimate of S.

Although windowing introduces bias into the spectrum, when carefully done it provides more good than harm, since we saw on page 73 that Ξ contains such large statistical fluctuations that distortions introduced by the window are usually quite negligible in comparison. In particular, if S is fairly smooth, then a moving average on it, which is S^w, would differ little from it. If S contains a set of sharp peaks, then S^w would differ greatly from S, but if the peaks are fairly far apart then they may still show up distinctly enough for the spectrum to be used effectively. Fig. 7.5 explains this more clearly. In actual computation, we have to trade off between variance and bias by varying the width of the window. The broader W is, the smaller is $V(\Xi^w)$, and the larger is $\beta(\Xi^w)$. We shall say more on this later.

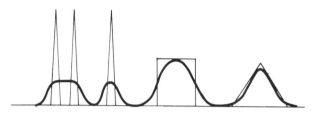

Fig. 7.5

Since Ξ^w is the Fourier transform of $w(t)\,\xi_a(t)$, we have

$$w(t)\,\xi_a(t) = \sum_i \Xi^w_{i/2T}\,\exp(cit).\qquad(19)$$

Putting $t = 0$ gives

$$\sum_i \Xi^w_{i/2T} = w(0)\,\xi_a(0) = w(0)\int_0^1 x(t)^2\,\mathrm{d}t.\qquad(20)$$

It is then clear that, provided we choose $w(0) = 1$, the total power in Ξ^w is equal to that in $x(t)$. Hence, application of a window on the autocorrelation preserves the total power of the spectrum. Note that $w(0) = 1$ implies $\sum \bar{W}_j = 1$ and $\sum \bar{W}_{j/2T} = 1$, which explains the choice of values for the Tukey and Daniall windows.

In Chapter 5 (page 56) we saw that the periodogram of white noise computed from N values contains $(\tfrac{1}{2}N + 1)$ independent random variables with constant mean value. These values correspond to *integer* frequencies. The elements for half-integer frequencies are arithmetically but not statistically independent from them. Equation (18) shows that the computed spectral element contains contributions from various frequencies, each given some weight. The weights become negligible as we move away from frequencies covered by the main lobe. Consequently, the width of the main lobe roughly determines the size of the set of independent variables over which averaging is performed. We shall call this the *spectral width* of the window.* The examples of the previous section show that a time window with cutoff $\pm T$ would have a main lobe that extends approximately from $(i/2T - 1/T)$ to $(i/2T + 1/T)$. (The rectangular window is only half as wide, but it is of no practical use. The Parzen window is wider, but not by much.) The smaller T is, the more independent elements we would have averaged over, and the smaller would be the variance of the final results. As the *arithmetically* independent elements are spaced at a frequency interval of ½ in the periodogram, a window with spectral width $2/T$ covers around $4/T$ elements. However, only half of them are *statistically* independent. The number of statistically independent elements covered by a window is half its degrees of freedom. Thus, a window with spectral width $2/T$ has approximately $4/T$ degrees of freedom.* This is because each value in the periodogram is the sum of squares of the real and imaginary parts of X, so that it has 2 degrees of freedom. Thus, although the number of degrees of freedom equals the number of arithmetically independent elements covered by the window, it is defined in terms of statistically independent elements.

We saw on page 65 that averaging over M independent variables gives an estimate with variance reduced by a factor of M. Hence, a spectrum of white noise computed using a window with cutoff T would have a variance of about $TS/4$. For others random processes the periodogram elements are not really independent. Consequently, the reduction in variance is not as large, though it is not very different. We also saw that the Parzen window has a broader main peak than Tukey or Bartlett windows for equal T, so that to get computed spectra with comparable variances T must be chosen somewhat larger (around 40%). Increasing T of course means that we have to compute more values of $\xi_a(t)$. Thus, there is a price to pay for the very desirable shape of the Parzen window.

The presence of the bias is due to the merging together of structures covered. A sharp peak with nothing on either side becomes a broad peak after windowing as the window mixes the peak with the 'nothing'. Two close sharp peaks would be merged into one peak. A rectangle with sharp edges will become something with sloping edges, and possibly with a flat top if the original rectangle is wide enough. In short, a window cannot reproduce anything narrower than itself. If we want to *separate* two peaks, i.e., reproduce the *gap* between the two peaks, we must use window narrower than the gap. If we want to see the shape of the peak itself, the window has to be narrower than the peak.

Thus, the amount of bias one can tolerate is very dependent on particular applications. For some, the shape of the peaks in the spectrum is important, so that very little bias can be accepted. For others, the positions of these peaks are sufficient information, so that the peaks need merely be resolved without being reproduced. In yet others, it does not matter if a cluster of peaks is merged into one as only the position of the whole cluster matters.

* Others have defined 'bandwidth' in an exact way. We do not require such exactness in practical applications. Same for degrees of freedom.

Generally speaking, to be usable a spectrum must have at least 16 degrees of freedom, preferably more. Such a window covers about 16 arithmetically independent elements. Any details we wish to see have to be at least that far apart. It often occurs that the spectrum contains too much detail for the width we have chosen. The remedy to take is not to increase T and narrow the window, since this will produce increased statistical fluctuations. Rather, we have to increase the number of independent spectral elements by taking a larger set of measurements. As we said on page 22, enlarging the interval over which $x(t)$ is measured from $[0, L]$ to $[0, 2L]$ permits us to produce a Fourier transform whose coefficients are placed at half the frequency interval. Peaks that were 16 elements apart now are 32 elements apart, so that a window covering 16 spectral elements will now resolve them. We shall comment on such considerations again in the next chapter.

Linear windows

So far, we have discussed windows that operate on quantities consisting of second-order terms in x, i.e., $w(t)$ is multiplied into $\xi_a(t)$ while $\bar{W}_j$ or $\bar{W}_{j/2T}$ operates on Ξ. Such windows are thus called *quadratic windows*. [$w(t)$ is also called a *lag window*.] At the moment this is rather out of fashion. Since FFT many became more interested in *linear windows*, also called *tapering windows*. These are multiplied into $x(t)$, so that the subsequent Fourier transformation produces X_i^w. Ξ_i^w is taken to be $|X_i^w|^2$. This produces a spectral estimate completely different from those seen before even if we use the same window. (In any case, we cannot use the same window, as shown below.) The advocates are convinced that this is the only way to handle leakage in the direct method for spectrum estimation. We think differently.

The thinking behind linear windowing is as follows. If we Fourier transform a sinusoid with a non-integer frequency, take $|X_i|^2$, and then apply, say, the Tukey window by computing

$$\tfrac{1}{4}|X_{i-1}|^2 + \tfrac{1}{2}|X_i|^2 + \tfrac{1}{2}|X_{i+1}|^2 , \tag{21}$$

for every i, we do not obtain a highly concentrated spectrum. Whereas if we apply the Tukey window to X_i, and then square, the result is much more concentrated. This leads to the claim that only the latter works for leakage correction. However, this is hardly a relevant example for us. Our purpose in spectrum estimation is to reproduce as closely as possible a given power spectrum. The relevant example to look at is an *autocorrelation* that is a pure sinusoid, corresponding to a power spectrum with one sharp peak. A pure sinusoid in $x(t)$ would often be removed as an additive periodicity in the pre-processing stage! In short, linear windowing certainly is useful if we want to compute X, but Ξ is quite a different matter. What we want is a good $\bar{W}(i-f)$ with a reasonably concentrated and smooth shape. For example, the Daniall windows, which we apply on $\Xi_{i/2}$, i.e., $|X_i|^2$, satisfy the criterion quite well. Our first complaint against linear windowing, is, therefore, that it is quite unnecessary. The job it is supposed to do is already being done adequately.

Our second complaint is that linear windowing does not reduce variance at all. This results from the need to preserve total power, which in quadratic windowing is achieved by choosing $w(0) = 1$ as shown in the last section. Here we require

$$\sum_i |X_i^w|^2 = \xi_a(0) = \int x(t)^2 \, dt . \tag{22}$$

The left-hand side is the same as $\int [x(t) \, w(t)]^2 \, dt$. The equation unfortunately does not lead to any simple condition on $w(t)$. Therefore, we instead require it to hold *on average*. This turns the left-hand side into $\int a(0) \, w(t)^2 \, dt$ and the right-hand side is simply $a(0)$. Thus, if $w(t)$ is chosen in such a way that

$$\int_0^1 w(t)^2 \, dt = 1 , \tag{23}$$

then, *on average*, Ξ^w contains the same power as $x(t)$. The power is not necessarily preserved for any particular $x(t)$! Equation (23) implies $\sum |W_j|^2 = 1$. Now let us look at the variance of X_i^w if x is white noise. We have

$$V(X_i^w) = \langle |X_i^w|^2 \rangle - \langle |X_i^w|^2 \rangle$$

$$= \langle \sum_{jj'} W_j X_{i-j} W_{j'}^* X_{i-j'}^* \rangle - \sum_{jj'} W_j W_{j'}^* \langle X_{i-j} \rangle \langle X_{i-j'}^* \rangle .$$

For white noise the Fourier coefficients have zero mean and different coefficients are uncorrelated. The second term vanishes and the first term contains only terms with $j = j'$. Thus we have

$$V(X_i^w) = V(X_i) \sum_j |W_j|^2 = V(X_i) .$$

In short, linear windowing alone does not reduce the amount of statistical fluctuation in Ξ, while quadratic windowing does. This difference is due to the restriction (23). In quadratic windowing we merely choose $w(0) = 1$ to preserve total power, and $w(t)$ for other t can be chosen as small as we please to reduce the weight given to the less reliable parts of $\xi_a(t)$. When we apply linear windowing on $x(t)$, $w(t)$ being smaller than 1 for some t must be compensated for by giving it values larger than 1 for other t in order to satisfy (23). Another way of putting it is that, every value in $x(t)$ is of comparable quality, and we cannot improve reliability by giving part of it less weight than others in computing the spectrum.

We already said that linear windowing only preserves power on average, not for any particular $x(t)$. A different way of looking at the problem is as follows: suppose $x(t)$ contains two sinusoids of approximately the same frequency and amplitude, then X^w would contain contribution from both as

$$X_i^w = \int W(i-f) \, X(f) \, df \sim X(f_1) + X(f_2) .$$

Now, $X(f)$ may be either positive or negative depending on the phases of the sinusoids. If the phases are such that $X(f_1)$ and $X(f_2)$ have the same sign, then the powers accumulate, and $|X^w|^2$ is about $4|X(f_1)|^2$. If $X(f_1)$ and $X(f_2)$ have opposite signs, they cancel and the power of X^w is almost zero. This cannot happen when we have quadratic windowing, because we are then averaging over a non-negative quantity Ξ. There can be no over-estimate of power through accumulation either as windowing is not followed by squaring.

It is by now clear that in linear windowing we cannot possibly use windows with large spectral widths. To start with, this would not cause any decrease in variance, whereas in quadratic windowing a larger window does have a positive effect. Second, with a wide window the chance of losing or over-estimating the power is increased as more components are added together. It follows therefore that windowing on $x(t)$ can at best have only a minor leakage correction effect because of our inability to use well-shaped windows like the Tukey or Parzen windows. These windows are far too wide. Instead, we are forced to use narrow ones like the cosine taper.

Finally, the relation between the linearly windowed spectral estimate and S is quite complex. This is different in quadratic windowing, in which we know that Ξ_i^w is an unbiased estimate of S^w, which is related by S by a convolution. In linear windowing, no such simple relation is available. This brings even more shakiness into the already shaky business of spectrum estimation. In short, our advice on linear windowing is 'don't'. Effective quadratic windowing methods are available for all possible cases.

Windows for segment averaging

Another FFT induced idea follows. Given N values of $x(t)$, we divide the whole into segments of M values each, perform a Fourier transformation over each segment, square the elements, and finally average these over all the segments. This can be considered to be equivalent to multiplying $\xi_a(t)$ by a window with cutoff $T = M/N$, and so would give a spectral estimate of comparable quality. Its advantage lies in computational simplicity, as it does away with the time consuming process of computing $\xi_a(t)$. At the same time, it requires FFT of several M-component vectors rather than that of one N-element vector for the purpose of computing $|X_i|^2$. This gives a slight decrease in computing time; but a far more important advantage is that when N is large an N-point FFT takes far too much core. Segmentation allows us to use much less core in transforming each segment separately. It is also more convenient to process data a block at a time in real-time operations.

These are valid points. Unfortunately people have been rushing to adopt the method without proper consideration of the quality of the resulting spectral estimates. Further, they have been using linear windowing on the segments, which we consider to be a waste of time. In this section we shall discuss the method to find the correct procedure.

Let us consider the segmentation of the interval $[0, 1]$ into sections of length T. For each segment, say the kth, we compute

$$X_i^k = \int_0^T \exp(-cit/T)\, x(kT+t)\, dt.$$

We can square this and then perform the same sort of manipulations as what we did on page 51, to produce

$$|X_i^k|^2 = \int_{-T}^T \exp(-cit/T)\, \xi^k(t)\, dt,$$

where

$$\xi^k(t) = \xi^k(-t) = \int_0^{T-t} x(kT+s+t)\, x(kT+s)\, ds, \quad \text{for } t \geqslant 0.$$

Summing over k gives our spectral estimate

$$\Xi_{i/2T} = \sum_k |X_i^k|^2 = \int_{-T}^T \exp(-cit/T)\, \xi_T(t)\, dt,$$

with

$$\xi_T(t) = \sum_k \xi^k(t).$$

In other words, the technique is equivalent to replacing the autocorrelation estimate $\xi_a(t)$ by $\xi_T(t)$ between cutoff points $\pm T$, with no windowing.

Obviously, the quality of such a spectrum depends on the quality of $\xi_T(t)$. Let us compare this with $\xi_a(t)$. Now $\xi_a(t)$ is $x(s)\,x(s+t)$ averaged over all possible values of s, from 0 to $(1-t)$. In comparison, $\xi_T(t)$ is $x(s)\,x(s+t)$ averaged over* $s = [0, T-t], [T, 2T-t], ..., [NT/M-T, 1-t]$. Obviously, $\xi_T(t)$ is likely to be less reliable than $\xi_a(t)$ because it includes fewer values, particularly for t close to T. It is also clear that, because $\xi_T(t)$ is unreliable for $t \sim T$, it is essential that $\xi_T(t)$ be multiplied by a window that decreases the contributions of such values towards Ξ_i. Since we never actually compute $\xi_T(t)$, this window has to be applied on the frequency domain, on the Fourier transform of $\xi_T(t)$, in other words $\Xi_{i/2T}$. This requires a window with a simple $\overline{W}_{j/2T}$, which means the Tukey windows. In short, in segment averaging, windowing is still possible, although rather less flexible.

In this way, we can still produce good spectral estimates, with leakage and variance under control. It is very curious indeed that the above described method has not been in common use. Instead, others have recommended linear windowing to suppress leakage. This, however, leaves untouched the problem of the unreliability of $\xi_T(t)$ for $t \sim T$. To solve that, it is suggested that we should use overlapping segments, e.g., in addition to the segments $[kT, (k+1)T]$ we should also Fourier transform $x(t)$ for segments $[(k+\frac{1}{2})T, (k+1\frac{1}{2})T]$ and include all $(2k-1)$ segments in

---

* Note that $\langle \xi_T(t) \rangle = (N/M)\, a(t)(T|t|) = a(t)(1 - |t|/T)$. It is thus closely related to the Bartlett window. In fact, Bartlett discovered his window through study of segmentation.

the final spectrum.   Now $\xi_T(t)$ is the average of $x(s)\,x(t+s)$ over
$s = [0,\,T{-}t]\,,\,[\tfrac{1}{2}T,\,1\tfrac{1}{2}T{-}t]\,,\,...,\,[(N/M{-}1\tfrac{1}{2})T,\,(N/M{-}\tfrac{1}{2})T{-}t]\,,\,[NT/M{-}T,\,1{-}t]\,.$
This does improve $\xi_T(t)$ for t close to T, but for small t there is negligible
improvement as the segments are not independent.   However, since the part
the method improves is precisely the part we can reduce by the use of a
window, the improvement is hardly worth while.   In short, the use of
quadratic windowing will correct leakage and improve stability quite well,
and there is no need for fancy methods.

8 Spectrum Estimation —A Summary

Preliminary steps

In the last chapter we covered all the theoretical background required for spectrum computation. It now remains only to put the previous discussions together and see how the knowledge is used in actual computation. In this chapter we summarize the practical aspects of spectrum estimation, starting with signal measurement, preprocessing and exploratory analysis. Summaries of different computing procedures are given. The next chapter will show computed examples.

In the signal measurement process, one starts with a signal source, and ends with a series of numbers that can be read by a computer. Many different things can go on between the two ends. It is beyond the scope of this book to study them, except in so far as these matters are affected by our computing requirements. Hence, we shall first look at the part of signal measurement that is closest to the computation, digitization, moving backwards to the more remote aspects.

Digitization is performed by A/D (analog-to-digital) converters. Such a device accepts an electric signal with a continually varying voltage, samples it at some interval, compares each sampled voltage with a reference voltage, and produces a number specifying the size of the sampled voltage relative to the reference. Two relevant considerations here are output word length and sampling rate. The cost of A/D converters increase rapidly with each. As we pointed out in Chapter 6 (page 69), if the values of $x(t)$ are used to compute ξ_a in the direct way, then 6 or 8 bits are quite adequate. If, however, the data are Fourier transformed, either to compute X or to compute ξ_a by the indirect method, then the word length should be larger.

In regard to sampling rate, we recall from Chapter 1 that if we wish to study the frequencies 0 to f, and we know that the incoming signal contains these frequencies *only*, then we would sample at $\Delta = 1/2f$. All the frequencies would be recovered quite accurately. (Those closest to f would be affected by low frequencies that have leaked out to frequencies above f.) If, however, the signal contains higher frequencies, then we have to sample at a smaller interval even if we are not interested in them. Thus, if the maximum frequency present is g, then we could either sample at $1/2g$ to recover all the frequencies, or we could sample at $(f+g)^{-1}$ to recover the frequencies 0 to f, with the frequencies g to $(f+g)/2$ being aliased and added to the range f to $(f+g)/2$. Instead of using a higher sampling rate one can remove the unwanted frequencies by filtering *before* digitization. This requires the use of *analog filters*. Whether we choose one alternative or the other is a matter of cost. If g is only slightly greater than f then an increase in the sampling rate may be a better proposition. If g considerably exceeds f analog filtering seems inevitable. Analog filtering also offers other potential benefits. For example, if we have an application in which the frequencies $[f, 2f]$ rather than $[0, f]$ are required, then we could use a filter that passes only the useful frequencies, and then sample at $\Delta = 1/2f$. This converts $[f, 2f]$ to $[f, o]$, and DFT of the data would recover the transform correctly. (Again note that because of leakage frequencies near f and $2f$ would be somewhat incorrect.) Another example is when we have a very wide-band signal covering $[0, nf]$ and we require all its frequencies. We could then use n filters and n A/D converters, each extracting out the range $[kf, (k+1)f]$, for $k = 0, 1, ..., (n-1)$. In fact, if the signal is very stable, so that we can obtain the same spectrum n times, then we could use one A/D converter n times for the n different bands. Finally, it is sometimes a good idea to remove or attenuate certain frequencies that are very much more prominent than others. For example, trends and periodicities can be removed this way provided we believe they are not physically meaningful. (Otherwise we should separate them digitally and analyse them.) Sometimes we wish to cut down sharp peaks in the spectrum so that there would be less leakage in the computation process. After the spectrum has been obtained we could then numerically 'undo' the filtering to find the true height of the peaks. (The idea of making the spectrum less lopsided and hence less likely to suffer from leakage is called *pre-whitening*.)

There are also instances when the sampling rate of the A/D converter is much higher than the input signal requires. In such cases it may be worthwhile to *multiplex*, i.e., to have several signals sharing the same A/D converter. A multiplexer simply switches one of the signals to the A/D converter for each sampling interval. It may also perform *de-multiplexing*, i.e., separating the numbers produced by the converter into several streams corresponding to different sources.

So far we have assumed electric signals, which need to be digitized. Some signals are produced in number form, e.g., stock market indices and population count. These need no digitization. Others are neither numerical nor electric, e.g., voice. These have to be converted to electric signals using various devices. Even electric ones may need amplification, noise removal, or level conversion. The last is particularly necessary as A/D converters require input signals to have a voltage within a standard range.

The instruments for converting non-electric signals to electric ones are generally called transducers. Transducers and other measuring devices all have certain ranges of frequency responses. They tend to cut out very low frequencies and also gradually attenuate high frequencies up to some maximum cutoff. These frequency ranges should be balanced with the sampling rates. Finally, we can easily compensate for the non-constant frequency response of our measuring devices by analog filtering or by numerical means. Thus, it is not always necessary to demand that the transducers have a flat response curve for all the frequencies we wish to study.

After digitization, one can either analyse the data as they are generated, so-called real time operation, or store them for subsequent processing. The former is essential for control applications, and may be desirable if the data are produced non-stop. Obviously, the processing facilities must be matched to the data rate. In real time operation the I/O capacity, the temporary storage, and the processing capability must be adequate. In off-line operation the main consideration is speed and capacity of recording facilities and portability and ease of conversion of data from one form to another. Paper tape may be a good way of recording small amounts of data, magnetic tape for large amounts. It is often possible to record on a slow medium high rate data produced in small and far-apart bursts if we buffer them properly. It may sometimes be sensible to record the analog data on a tape recorder for subsequent digitization. For some loss of fidelity, we save the cost of duplication in A/D converters. These are some of the many possible ideas one should consider.

Once the signals have been digitized, we are ready to perform preprocessing. This is again very dependent on the particular problem and instrumentation. Roughly, there are three different types of operations. The first is formatting, i.e., the conversion of data from the form in which they are produced to a form suitable for analysis. An obvious example is conversion between different word lengths. Or, the values may need to be converted into floating point representation. As A/D converter output is in terms of the reference voltage, it has to be scaled back into a physically meaningful unit. Some blocking and unblocking may also be necessary. All these are trivial in principle, but they must be correctly performed before analysis can start.

The second type is called data editing, during which a human examines the data, corrects any errors found, and also performs some simple operations on them. One of the important purposes is to eliminate the 'outliers' or 'wild points', values that are grossly different from what they should be. Such points are generated by instrumental faults of various kinds, particularly in data storage, retrieval or transmission. Thus, a speck of dust can completely wipe out a number stored on a magnetic tape, so that when the tape is read back a anomalous number results. The outliers are extremely damaging for any analysis that involves least squares approximation, which spectral analysis does, because a large error is magnified by squaring, so that the approximation one gets tries hard to reduce the error of the wild point, at the expense of the good points. There are what one calls robust estimation methods that are relatively insensitive to such points, but these are costly in computer time, and it is usually easier to eliminate outliers by human intervention. Usually, these values are so grossly in error that their presence can be immediately detected by inspection. With the ready access to interactive terminals most users enjoy these days, the task is far from a difficult one in any case.

The third type of preprocessing concerns in general the statistical properties of the signal. The data are examined for the presence of trends and periodicities, and appropriate action is taken to remove them as described in Chapter 4. As we said there, the methods are not foolproof and should be applied with care. We must always have a clear idea of the reason for non-stationarities.

Many authors have recommended that even for apparently time-invariant signals we should always remove the mean, i.e., subtract from $x(t)$ the time average $\int x(t)\,dt$. The motivation here is to produce a zero-mean random process. In our opinion, this practice, though usually helpful, should nevertheless be done with caution. To put it simply, a zero mean process does not necessarily have a zero *measured* mean. In fact, $\int x(t)\,dt = X_0$ is a random variable with zero (theoretical) mean and variance S_0. Forcing X_0 to zero, and thus forcing S_0 to zero, creates an artificial trough in the spectrum if S for other low frequencies do not happen to be zero. This sharp trough, like a sharp peak, leaks out, disturbing the computed spectrum. Such an undesirable effect has been observed by, for example, Hannan (1960, page 73). Therefore, mean removal should be performed only if X_0^2 considerably exceeds other components, which would show up in $x(t)$ being predominantly above or below zero throughout the interval $[0, 1]$.

In linear trend removal we subtract $\alpha_0 + \alpha_1 t$ from $x(t)$. This always removes the mean. However, as it also removes much of the low frequency components, no sharp trough is created, and there is no serious effect.

We mentioned trend and periodicity removal by digital filtering in Chapter 4, and earlier we mentioned analog filtering for the same purpose. When should we use which? Generally speaking, digital filters are much more flexible. We can implement almost any desirable filter with relative ease, and they can be changed simply by adjusting a few parameters or by writing a new subroutine. We can filter the same data repeatedly using different filters. This is particularly important when the data contain parts that have to be analysed separately. However, there is one thing digital filters cannot do: it cannot eliminate aliasing, which has to be controlled by analog filtering before sampling and digitization.

Trend and periodicity removal by digital filtering is only part of a larger problem of separating the spectrum into parts with different properties.

For example, we may be interested in the frequencies 0 to f, but the range 0 to $0.1f$ may be particularly important. Overall, we would like to look at spectral elements ϕ frequencies apart, but within the smaller range we wish to space them 0.1ϕ apart. Since we wish to see maximum frequency f, the signal is analog filtered to remove higher frequencies, and then sampled at, say, $\Delta = 0.4/f$ until we have sufficient data for looking at the *whole* spectrum at 0.1ϕ spacing. From page 21 we know that data measured over an interval of length L permits us to look at frequencies $1/L$ apart. However, because we use window with cutoff $\pm T$ the actual interval length over which Fourier transformation is performed is only $2TL$, so that $0.1\phi = 1/(2TL)$ means $L = 5/(T\phi)$. At a sampling interval of $0.4/f$, we need $N = L/\Delta = 12.5f/(T\phi)$ values of input data. For the whole spectrum a spacing of ϕ is fine enough, so we actually use only 10% of the data. To obtain the smaller spectrum, we perform digitally a low pass filtering on the complete data set to remove all frequencies above $0.1f$. Now we can sample at an interval 10 times larger, or, we take from the filtered data one number out of every ten. This allows us to look at a maximum 10 times smaller, but because the data are taken over the long interval, we can space the frequencies ten times denser. Thus, by the use of digital filtering, we manage to look at all the details we require by processing only 10% of the data at a time: we obtain a wide but coarsely spaced spectrum by using values spaced closely over a small interval, and a shorter but more finely spaced spectrum using filtered data over a large interval but spaced farther apart.

The above example also illustrates some considerations we must make about computation parameters. We showed that the number of measured values of x we require is at least

$$N = L/\Delta = 1/(2T\phi\Delta) = f/(T\phi) ,$$

as we require $L = 1/(2T\phi)$ and $\Delta = 1/2f$. We see that N increases if T or ϕ decreases or if f increases. In other words, the wider or denser is the spectrum, or the smaller is the cutoff (smaller cutoff gives increased stability as discussed on page 85), the more data we have to take. Note that f determines Δ; T has to be reasonably small, say <0.2; T and ϕ determine L; and L and Δ together determine N. Note that f/ϕ is the number of points on a spectrum, in other words M. Thus $N = f/T\phi$ means that $NT = M$, as it should. (See page 55.)

The above values are the *computation parameters* of a spectrum. Choosing T, ϕ, or Δ too small is wasteful. Making them too large produces poor results. Unfortunately, we do not usually know beforehand what their values should be for certain. This is why we must do some exploratory work, which is described in the next section.

The exploratory stage

As the last section shows, how we are to carry out the measurement of the signal is very dependent on the properties of the spectrum we wish to compute. Values of Δ and T must be small enough, and N large enough, to obtain reliable spectra with good resolution of details. Of course, when embarking on a new spectral analysis problem we do not have clear idea about the properties of the spectrum. The exploratory stage is for the purpose of getting some idea of these. By making some guesses about the spectrum and computing preliminary spectra based on these guesses, we gradually move towards a more accurate determination of the required properties. After these have been determined we will be in a position to process additional data from the same source or similar sources using the information already derived.

The computation process followed during the exploratory stage is often different from the procedure adopted for subsequent data processing. In the latter, we want to compute the final results at least cost. In the former, we would often obtain important insights about the spectrum from intermediate results. Consequently, we would make use of techniques that produce such information even if they may not be the most efficient.

Since we know that our measuring devices would always have limited frequency responses, this gives us an initial guess for f. However, the signal may actually have a smaller frequency range, and in any case we may find that part of the frequencies provide no new information and so can be safely eliminated. These questions are fairly easy to settle, perhaps simply by looking at the Fourier transform of some data. In short, choosing Δ is not difficult.

The choice of a suitable window width and the determination of the required resolution ϕ is rather more difficult. In any computed spectrum we have some bias introduced by the window, and random fluctuations which windows are supposed to suppress but never do completely. When we compare spectra computed with different N, T, etc., it is not always clear whether the differences we see are fluctuations or genuine new features shown up by a better window. In this we must rely heavily on what we know about the nature of the signal to decide which spectrum makes more sense. The procedure usually followed is to choose an N and try a few windows with varying T, and decide by this if N has been well chosen. We would of course like T to be as large as practicable and to keep N small to save measurement and computation costs.

As discussed in the last chapter, we can obtain a stable spectral estimate either by performing windowing on the autocorrelation, or by averaging the periodogram, or by segment averaging. In general, the first is the most suitable for exploratory studies. There are several reasons. First, as an intermediate result the autocorrelation provides a great deal of information about the spectrum. For example, if $\xi_a(t)$ has a sharp peak at $t = 0$ that dominates the rest of the values, we can expect the computed spectrum to be relatively smooth. Whilst there can still be peaks in the spectrum, they would have broad bases or a slowly varying background. (See, for example, the two figures of Chapter 6.) Isolated peaks in the spectrum would show up as oscillations with corresponding frequencies in the autocorrelation. Second, it is easy to experiment with windows of different shapes and cut-offs when we have $\xi_a(t)$. In fact, it is always useful to multiply $\xi_a(t)$ by a series of windows with cutoffs varying, say, from 0.05 to 0.2, and compute a set of preliminary spectra. Comparing these together gives us a fairly good idea whether N is large enough. If it is too small, then there is much crowding together of details in the spectrum. The computed spectrum for $T \sim 0.05$ would have too much bias, while that for $T \sim 0.2$ would suffer from excessive variance. The few preliminary spectra would differ widely from each other, indicating that ϕ is actually smaller than we have guessed and that N should be chosen larger. But if N is already large enough, then the computed spectra for different T would show similar structures, those with small T being somewhat more diffused. If we find that parts of the spectrum are not as well resolved as the others, then we might try isolating the high resolution part by filtering and analysing it separately, as shown in the last section.

It needs pointing out that there is not much use in computing $\xi_a(t)$ for t appreciably larger than 0.2, as such values are extremely unreliable and hence quite uninformative.

The periodogram, as an intermediary, is also quite informative. Where S is large, Ξ tends to be large and to have more fluctuations. Thus, a visual inspection of Ξ tells us much about S. It is also possible to experiment with different window widths by choosing different values of m, say from 5 to 20 for Daniall windows. However, for exploratory studies the periodogram suffers the disadvantage of containing too many numbers (N), whereas we need to obtain only a small part of ξ_a in the autocorrelation method.

Segment averaging is the least satisfactory technique for exploratory studies. It gives us no useful intermediary results, and to experiment with different values of T one has to repeat the whole calculation. Further, in theory at least the technique produces spectral estimates of lower quality than the others as shown on page 89.

Choice of computing procedures

We now turn to what one might call the 'production' stage of spectral analysis, when we know what appropriate values of Δ, T, N, etc., to use. The question is now to choose the most efficient computing procedure. In this section we recount all the available techniques and point out under what circumstances each is the most suitable.

1. *The autocorrelation method.* This has the advantage of producing ξ_a as an intermediary result, and is the obvious one for applications in which both the autocorrelation and the spectrum are useful. The method involves two major steps: (1) computation of $\xi_a(t)$ up to $t = T$, and (2) Fourier transformation with windowing. For each step we have a choice of two alternatives, (a) and (b):

($1a$) Direct computation of $\xi_a(t)$ as $\int_0^{1-t} x(s)\,x(s+t)\,\mathrm{d}s$. Use this if TN (which is called M) is small or moderately large, say less than 256.

($1b$) Compute $\xi_a(t)$ by Fourier transform. Use this if M is fairly large. Again we have two choices. If x is short enough to be contained completely in core, then just Fourier transform x plus M zeros. If N is very large, segmentation is necessary. See page 69 for details.

After obtaining $\xi_a(t)$ we divide $\xi_a(0)$ by 2 to produce $\xi_a'(t)$, and then:

($2a$) Directly compute

$$\Xi_{i/2T}^{w} = \frac{1}{M} \sum_{j=0}^{M-1} \cos(\pi ij/M)\, \xi_a'(j/M)\, w(j/M) .$$

Use this if M is small, say not more than 64.

($2b$) Attach M zeros to $\xi_a'(j/M)\, w(j/M)$ and perform a $2M$-point FFT on the enlarged vector. Discard the second half of the results. Take the real part of the first half. This approach is for large or moderately large M.

As an alternative to (2), leave out $w(j/M)$ from both equations, but window after Fourier transformation by the following:

$$\Xi_{i/2T}^{w} = \tfrac{1}{2}\Xi_{i/2T} + \tfrac{1}{4}(\Xi_{(i-1)/2T} + \Xi_{(i+1)/2T}) . \tag{1}$$

For $i = 0$ or M, remember $\Xi_i = \Xi_{-i} = \Xi_{N-i}$, so that

$$\Xi_{0/2T}^{w} = \tfrac{1}{2}\Xi_{0/2T} + \tfrac{1}{2}\Xi_{1/2T} \qquad \text{and} \qquad \Xi_{M/2T}^{w} = \tfrac{1}{2}\Xi_{M/2T} + \tfrac{1}{2}\Xi_{(M-1)/2T} .$$

Already, we see how confusing the picture can be to a beginner. He is faced with three or four choices at various stages. If one wants to be fastidious, one can distinguish between perhaps thirty different computing procedures, by taking different combinations of the available alternatives. Yet, there is nothing so very difficult about the computation. Here lies the source of much trouble about spectrum estimation: it is easy to *do* but often confusing to *think about*, with the result that people often do not bother with the latter. They hold to the familiar techniques without carefully considering alternatives.

2. *The smoothed periodogram method.* This has the advantage of producing X as intermediary result. It becomes difficult to use if N is very large because of its core requirement. Again there are two steps: attach N zeros to x, perform a $2N$-point FFT and square to produce $\Xi_{i/2}$, $i = 0, 1$, ..., N. (The other half are redundant.) Apply one of the Daniall windows with appropriate m. Each value $\Xi_{i/2}^{w}$ is the average of $2m+1$ values of $\Xi_{i/2}$. We would take i at $0, m, 2m, ..., N$, producing an N/m point spectrum. Fortunately, this method allows few alternatives and so is not too confusing.

3. *Segment averaging.* Computationally this is the most efficient. It does not require ξ_a, nor does it take much core. However, no useful intermediary results are produced. It is good if we want the spectrum and nothing else.

The time and core requirements are smallest if we take segments of length T. Thus, each segment has M values of x. We attach M zeros to each segment, perform a $2M$-point FFT, square, and then average over the segments to produce $\Xi_{i/2T}$. Then we apply the Tukey window according to (1).

If M is quite small, say less than 64, then there is no need to use FFT. For each segment, compute the real part of each coefficient as $R_{i/2T} = \Sigma \cos(\pi ij/M) x(kT+j/M)$ and the imaginary part as $I_{i/2T} = \Sigma \sin(\pi ij/M) x(kT+j/M)$. $\Xi_{i/2T}$ is $(R_{i/2T})^2 + (I_{i/2T})^2$ summed over all the segments and divided by N. Windowing follows.

We can, however, choose a combination of methods 2 and 3. For example, we can take segments of $2M$ values each, perform $4M$-point FFTs, average over segments, and then apply the modified Daniall window with $m = 2$. Initially we get a $(2M+1)$ point periodogram from each segment. After use of the window we have $(2M+1)/m \sim M$ points in the computed spectrum, which is just what we want. In theory, the quality of spectral estimates improve with segment size, but this does not appear to be very significant in practice. In any case, we still get no useful intermediary results. So perhaps the extra work required here is not worth while.

Method 2 is to be preferred when we require both X and ξ_a as intermediary results. We already have X. Inverse transforming $|X_{i/2}|^2$ gives the autocorrelation as shown on page 53. In actual computation it works out as follows:

$$\xi_a(j/N) = \sum_{i=0}^{2N-1} \Xi_{i/2} \exp(\tfrac{1}{2}cij/N) , \tag{2}$$

a $2N$-point inverse DFT as discussed on page 31. Because of aliasing this is the same as

$$\xi_a(j/N) = \sum_{i=-N+1}^{N} \Xi_{i/2} \exp(\tfrac{1}{2}cij/N) ,$$

and as $\Xi_{i/2} = \Xi_{-i/2}$, we also have

$$\xi_a(j/N) = 2 \sum_{i=0}^{N} \Xi'_{i/2} \cos(\pi ij/N) , \tag{3}$$

where $\Xi' = \Xi$ except that $\Xi'_0 = \tfrac{1}{2}\Xi_0$ and $\Xi'_{\frac{1}{2}N} = \tfrac{1}{2}\Xi_{\frac{1}{2}N}$. These two terms have to be halved because they have no negative index counterparts. Sometimes we wish to remove the mean of $x(t)$. (See page 94 above.) This can be achieved simply by putting $\Xi_0 = 0$. In any case, another way of writing (3) is

$$\xi_a(j/N) = 2\,\mathrm{Re}\left(\sum_{i=0}^{N} \Xi'_{i/2} \exp(cij/2N) \right). \tag{4}$$

Either (2) or (4) can be computed by a $2N$-point FFT. This is efficient if N is fairly large. If N is quite small then (3) is more convenient to use.

Estimation of the cross spectrum*

We have been looking at the evaluation of $S_i^x = \langle X_i^* X_i \rangle$. Now we wish to study the estimation of $S_i^{xy} = \langle Y_i^* X_i \rangle$. There is no essential difference between the two problems. As before, we have the choice of directly evaluating $X_i Y_i^*$ and then computing moving averages or segment averages, or evaluating the cross-correlation $\langle x(s+t) y(s) \rangle$ first followed by windowing and Fourier transformation. When the data vectors are long and segmentation is necessary, and we also happen to require both S^{xy} and the cross-correlation, we would use the segment correlation method given on page 71. If we need only the former, then just averaging $Y_i^* X_i$ over the segments and windowing are sufficient.

One important difference is that the cross-correlation is not symmetric, $g(-t) \neq g(t)$ in general. Consequently, no simplification along the line of equation (6) in Chapter 5 is possible. If we denote the estimate of $g(t)$ as $\xi_g(t)$, we would compute estimated values of G_i by

$$\int_{-T}^{T} \exp(-cit/2T)\, \xi_g(t)\, w(t)\, \mathrm{d}t . \tag{5}$$

This cannot be converted into a Fourier transformation over $[0, T]$. If sampled values are used, we have

$$G_i \sim \sum_{j=-M}^{M} \exp(-cij/2M)\, \xi_g(j/N)\, w(j/N) . \tag{6}$$

To put this into standard form of DFT, we would have

$$\sum_{j=0}^{2M-1} \exp(-cij/2M)\, \xi_g'(j/N)\, w(j/N) ,$$

where
$$\xi_g'(j/N) = \xi_g(j/N) , \qquad 0 \leqslant j \leqslant M-1 ,$$
$$= \xi_g(M/N) + \xi_g(-M/N) , \qquad j = M ,$$
$$= \xi_g[(j-2M)/N] , \qquad M+1 \leqslant j \leqslant 2M-1 .$$

The inequality of $g(t)$ and $g(-t)$ makes no difference to the direct methods as far as computation procedures are concerned. However, whereas $X_i^* X_i$ is always real $Y_i^* X_i$ is in general complex.

Generally speaking, the quality of cross-spectral estimates is poorer than power spectrum estimates of equal degrees of freedom. The phase is extremely difficult to estimate accurately. An indication of this difficulty is as follows. Suppose we perform no averaging of any kind at all; that is we take simply

$$S_i^x \sim X_i X_i^*, \qquad S_i^y \sim Y_i Y_i^* \qquad \text{and} \qquad S_i^{xy} \sim X_i Y_i^* ,$$

so that we would have $\kappa_i \sim |S_i^{xy}|^2/(S_i^x S_i^y) = 1$.

* This may be omitted on a first reading.

In other words, if there is an insufficient number of degrees of freedom the coherency spectrum would be approximately 1. A result like this is extremely misleading.

Cross-spectral estimates often become much better if we 'align' the data, i.e., shift $\xi_g(t)$ such that its maximum falls on $t = 0$. We shall not go into a detailed discussion about why this occurs, but the following will give an indication. Suppose $\xi_g(t)$ has been shifted by Δ, so that we are really trying to estimate

$$G_i' = \int g(t + \Delta) \exp(-cit) \, dt = \exp(ci\Delta)G_i \, .$$

By windowing, we now average over values of G'. Now, as $g(t + \Delta)$ has its maximum at $t = 0$, it comes close to being symmetric, so that G' comes close to being real and positive like S. When values of G' are averaged together, we do not get the amount of cancellation we would when averaging over values of G. This is why the results are more reliable. Having estimated G', we then divide the result by $\exp(ci\Delta)$ to obtain an estimate for G_i. This amounts to a phase change only. If we want only κ_i then we can simply put the value of G_i' into equation (18) in Chapter 5.

The above works when we estimate G via g. What if we are using the direct method? As we saw, alignment amounts to changing the phase of G_i by $2\pi i\Delta$, such that G_i' comes as close as possible to being real. The equivalent requirement is to find Δ such that the phase of $X_i Y_i^* \exp(ci\Delta)$ fluctuates randomly about 0, or, that $\phi_i + 2i\Delta$ should fluctuate randomly about 0. This is done by fitting a straight line passing through the origin to the phases of $X_i Y_i^*$. The slope of this line is $-2\pi\Delta$. We then compute moving averages of $X_i Y_i^* \exp(ci\Delta)$ as estimates for G'. Finally G_i is obtained by taking $G_i' \exp(-ci\Delta)$. Unfortunately, alignment cannot be carried out in the segment averaging method.

Because of the poorer reliability of cross-spectral estimates the need for exploratory studies becomes even more imperative. Above all, the coherency spectrum should be carefully examined to make sure that it makes sense physically. In addition to harder computation, cross spectra are also more difficult to interpret. One example of the use of cross spectrum is shown on page 113. In view of the need for fairly extended discussion, the reader should consult other books before starting. Jenkins and Watts is recommended for a good discussion of the computation of cross-spectrum and its use.

9 Spectrum Estimation —Examples

Artificial data

In this chapter we illustrate the discussion of Chapter 8 with computed examples. We shall first look at artificially generated data with known theoretical spectra so that the performance of different computing procedures may be compared. Then we shall see the use of spectral analysis in the study of turbulent fluid flows. While in the former example we shall stress the computational aspects, in the latter we shall emphasize preprocessing and spectrum interpretation.

Our artificial data are produced by first generating a set of random numbers and then filtering them in some way. Random number generator subroutines are available on most computers with standard mathematical packages. They usually generate uncorrelated numbers uniformly distributed between 0 and 1. For our purposes it is best to produce Gaussian random numbers with zero mean and unit standard deviation. These can be obtained in a reasonably satisfactory way by adding 12 uniformly distributed numbers and subtracting 6. In our particular case a fast subroutine for Gaussian numbers was provided to us by R. Brent of the Australian National University. (Details of this routine are given in *Communications of the A.C.M.*, 1974, **17**, 704-6.)

We take $N = 512$ and generate a set of 512 random numbers. We shall call them n_j. Then we produce two pieces of artificial data by computing

(1): $x_j = \frac{1}{4}n_{j-1} + \frac{1}{2}n_j + \frac{1}{4}n_{j+1}$, for $j = 1, 2, ..., 510$;

$\quad x_0 = \frac{3}{4}n_0 + \frac{1}{4}n_1$; $\quad x_{511} = \frac{3}{4}n_{511} + \frac{1}{4}n_{510}$;

and

(2): $x_0 = n_0$; $x_1 = n_1$; $x_j = x_{j-1} - \frac{1}{2}x_{j-2} + n_j$, for $j = 2, 3, ..., 511$.

(Statisticians call the former a moving average process, and the latter is an autoregressive process. Engineers will recognize them as non-recursive and recursive filtering.) The two random processes have the following theoretical spectra:

(1): $S_i = [1 + \cos(2\pi i/N)]/2$

(2): $S_i = [2\tfrac{1}{4} - 3\cos(2\pi i/N) + \cos(4\pi i/N)]^{-1}$.

They may be derived easily by the methods shown in Chapter 11.

Since the data are artificial, we could assume that they have been measured over any interval e.g., $[0, L]$. Each point in the spectrum then corresponds to frequency i/L. The sampling interval is $\Delta = L/N$, as we have N values of x over $[0, L]$. The maximum frequency we can look at is then $1/2\Delta = \tfrac{1}{2}N/L$. Thus, we could take i between 0 and $\tfrac{1}{2}N$. However, we would average over groups of frequencies to obtain stable estimates. Hence, we choose a truncation point T for the correlation method, a segment length T for the segment averaging method, and a window width by picking a suitable value of m for the frequency averaging method. As we remarked, m is equivalent to $1/T$. Let us choose $T = 1/16$ or $m = 16$. Or, we assume that the important structures in S are at least $8/L$ integer frequencies apart. ($\phi = 1/2TL$.) Since $M = TN = 32$, we compute spectra of 32 (or perhaps 33) values each. This of course agrees with the number we get by dividing the total width of the spectrum, $\tfrac{1}{2}N/L$, by the separation between different spectral elements, $8/L$. The computed values correspond to the power at frequencies $i/2TL = 8i/L$ ($i = 0, 1, ..., 31$).

Again we remind the reader that the value of L and the physical frequencies $i/2TL$ do not appear in our actual computation. Computing DFT automatically puts the correct frequencies in. It only requires us to identify each value in the computed spectrum with the correct frequency after we have obtained it.

We show in Figs 9.1 and 9.2 twelve spectra computed from the two pieces of data using six different methods for each piece. In each a is the spectrum computed by simply truncating the autocorrelation at $\pm T$ and performing DFT; b is obtained by applying the Tukey window with $\alpha = \tfrac{1}{2}$ to a; c is produced by frequency averaging with equal weight for each value $\Xi_{i/2}$ in the periodogram; d is produced by segment averaging using no window; e by segment averaging using a linear window before Fourier transformation; f is d after smoothing with the Tukey window (page 98).

There seems little doubt that b, c and f, all being spectra that have been subjected to quadratic windowing of some sort, are better than those that have not been. The three good ones come close to actually reproducing the shape of the theoretical spectrum in each case, and they have comparable quality though the exact shapes are not all the same. In e, the linearly windowed spectrum, is slightly smoother than the two unwindowed ones, but it is certainly inferior to the quadratically windowed spectra.

Because both theoretical spectra are quite smooth and simple in structure, the computed results reproduce them reasonably well. If the data have a more complex theoretical spectrum the results tend to be much less

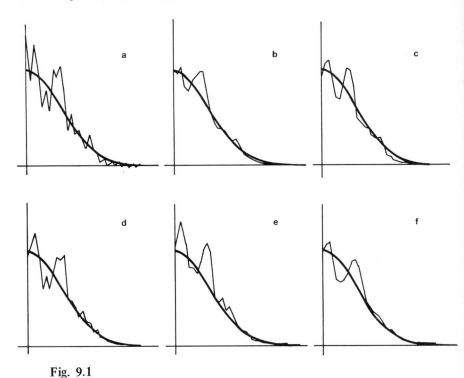

Fig. 9.1

satisfactory. In particular, details that are less than 8 integer frequencies apart would not be resolved at all. (Even this is a bit too optimistic. In fact, two sharp peaks 8 integer frequencies apart would not be resolved either. They have to be 16 integer frequencies apart to assure separation. The Tukey window, for example, has a main lobe covering $4/T = 64$ half-integer frequencies.)

Turbulence data analysis

The investigation of turbulent fluid flows is an area which can make good use of spectral analysis. Turbulent flows occur often in our experience. For example, the stirring of cream in a coffee cup, the gusting of the wind, the cauliflower shape of a cumulus cloud, or the roar of a jet engine, are all turbulent flows. To put it simply, turbulence is a random fluid motion, sometimes superimposed on an otherwise steady flow, which occurs if layers of a fluid slide or shear across one another. Viscous forces in the fluid, interacting with the fluid momentum, cause an otherwise steady flow to break up into random eddies, a kind of rotational flows. This can occur close to a solid surface, which slows down the flow of the fluid near it and hence produces layers of different flow speeds, or in a wake or jet where flows of different speeds meet. The equations governing the motion

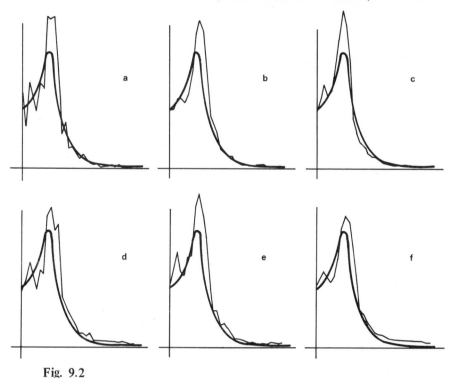

Fig. 9.2

in turbulent flows are non-linear partial differential equations, which can only be solved in terms of initial values. To solve the equations for even simple flows require enormous amounts of computation, even if it were possible to obtain all the initial values required. In view of the difficulty of analytic solutions, it is more usual to try to build an empirical picture of a turbulent flow in terms of some of its statistical properties.

In turbulence analysis, the random process usually analysed is the velocity of fluid flow at some point. Besides the computation of a spectrum, the measured values are also used in many other types of analysis. As a result, we usually require both X and $\xi_a(t)$, and the smoothed periodogram method is preferred, with $\xi_a(t)$ being obtained usually from Ξ; X is saved for further analysis.

The basic measuring instrument in turbulence research is the hot wire anemometer, which consists of a very fine, electrically heated wire, typically 3 μm in diameter and 1 mm in length. It is supported between much thicker wires at the tip of a measuring probe inserted into the flow. The instantaneous heat loss from the wire is determined electrically, and gives a measure of the flow speed around the wire. (This is wind chill.) Because the thermal inertia of the wire is very small, the instrument can respond to fluctuations in fluid velocity of the order of thousands of hertz (cycles per second). In a typical low speed laboratory boundary layer it is found that, although most of the energy of the fluctuation is at frequencies of the order of hundreds of hertz, there is a long tail to the spectrum which is non-zero to at least 50 kHz. It is in this tail that kinetic energy of the turbulent flow is being dissipated, by viscous action, into heat. If we confine our interest for the moment to that part of the spectrum which contains something like 95% of the energy, then we are thinking of a measurement bandwidth of a few thousand hertz, and thus a sampling frequency double this. However, the fluctuating velocity of the flow have different behaviour at different points. We must often sample simultaneously at more than one point. At each point we may require two velocity components, obtained with two hot wires at right angle to each other. In some cases non-velocity quantities, such as temperature, or fluid density in the case of mixed fluid flow, may be measured also. And finally, to obtain statistically stable results the data collection time must be many times the period of the lowest frequency of interest. That is to say, we require L to be large enough so that $1/(2TL)$ is less than ϕ, which may be of the order of 1 Hz. All this means is that the total data gathered in one experiment may be hundreds of thousands of values, as $N = f/T\phi$. For example, a rather modest one at that, with $f = 1000$ Hz, $T = 0.1$, $\phi = 1$ Hz, we would have $N = 10^4$.

In designing an experiment it is first essential to reduce the number of measurements to a minimum, in order to limit the data 'explosion' seen above. Second, the bandwidth f should be selected with care, also to limit the total data without loss of useful information. For these reasons, exploratory experiments, over a small number of points with large bandwidth and over a larger number of points with smaller bandwidth, are good practice. Recording of measured data, either by analog or digital means, is useful to allow different analyses to be carried out repeatedly with no need to re-run the experiment. Also, it is good to compare the results of different computing procedures using the same data. Before sampling and digitization it is essential that unwanted frequencies are eliminated from the signal to prevent aliasing, and for this purpose a good quality low-pass or band-pass filter is needed. When a number of analog signals are to be digitized at once, each must be filtered independently before sampling, requiring a separate filter for each data channel. If cross correlations and cross spectra are of interest, the filter characteristics of the different filters should be matched as closely as possible. It may also be considered desirable to use analog sample-and-hold circuitry to sample all signals at the same instant so as to eliminate the effective digital phase shifts due to different sampling instants. All analog filters will introduce phase shifts near the cut-off frequency, the higher the filter order the greater the possible phase shifts. Finally, remember that f has to be somewhat smaller than $1/2\Delta$ to ensure that no serious aliasing due to leakage occurs.

Having digitized and recorded his data, the turbulence researcher probably first preprocesses the data using anemometer calibration parameters and a combination of values from different sources to obtain component values.

Mean values may be determined and used to non-dimensionalize the data (i.e., divide each value by the mean) and also to remove the mean. Mean removal is usually helpful here in that it eliminates the steady flow of the fluid, leaving behind only the random fluctuations. Division by the mean then measures the ratio of fluctuations to the steady flow, giving what we call the turbulent velocity excursions, a zero-mean process. The mean square of the fluctuations may also be determined as a measure of the turbulent energy.

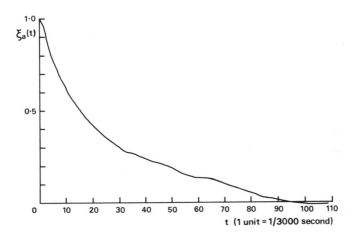

Fig. 9.3

Fig. 9.3 shows an example of the autocorrelation of turbulent velocities. This can be computed either directly, or, more often, by the double use of Fourier transformation. It is used to give an indication of the scale of coherent eddies in the turbulence. The value of t at which $\xi_a(t)$ crosses the t-axis and becomes negative gives an indication of the rotational flows, as it shows that $x(s)$ and $x(s+t)$ tend to have opposite signs, being opposite sides of such flows reversals. Multiplying t by the mean velocity then gives an indication of the size of the eddies. By comparing $\xi_a(t)$ obtained from different points in the fluid flow, we see how the structure of the flow is changing. For example, the flow data obtained from the front and back of an object placed in the flow show how its shape affects the motion of the fluid. Such results can then be used in, say, aerodynamic designs.

Using the FFT algorithm we obtain the periodogram, on which we perform smoothing to obtain a power spectrum like that shown in Fig. 9.4. This again allows the scale of eddies to be compared from one flow region to another. High frequencies of course indicate small eddies, and low frequencies large ones. In most turbulent flows the power spectrum is only slowly changing with frequency. (Assuming, that is, we have done enough smoothing to make the results stable.)

Turbulence may be thought of as an energy transfer process, with the 'orderly' kinetic energy of the steady flow being converted to the turbulent kinetic energy of eddies of decreasing sizes, to the final random molecular kinetic energy of the fluid molecules, which is just heat.   If we plot the power spectrum of turbulent velocity multiplied by various powers of the frequency, we have the spectrum of the various time differentials of turbulent energy.   For example, Fig. 9.5 shows the so-called dissipation spectrum where the square of the frequency has been multiplied into the power spectrum;   this represents turbulent energy dissipation at various frequencies and the maximum indicates frequency scales where dissipation is proceeding most rapidly, which is another useful parameter for comparison between flows.

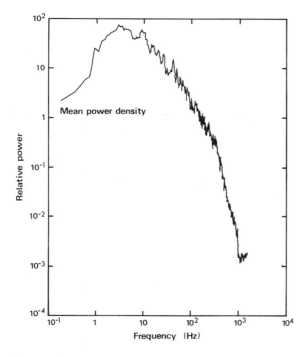

Fig. 9.4

Cross spectra provide a means for determining the scale of coherent eddies between two points in a flow, or between component velocities or other turbulent parameters at one or more points.   If two points have some degree of coherence, then a cross correlation would have a maximum value at a time delay approximately equal to the distance between the two points multiplied into some flow speed.

In summary, the turbulence researcher uses spectral analysis as one of the methods for understanding the underlying process causing, maintaining and dissipating a turbulent flow.   To this end, and in common with other users of spectral analysis, he must be prepared, firstly, to artificially limit his signal bandwidth to prevent aliasing, and secondly, to process very large amounts of data to achieve stable spectra.   (Leakage is not a serious problem because the power spectrum is usually smooth.)   His experiments should include some exploratory work so as to determine the best experiments for the major effort in view of the huge amounts of data which

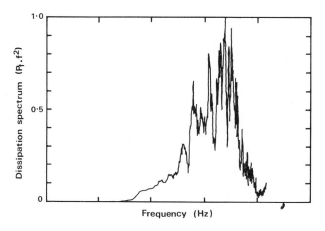

Fig. 9.5

could be gathered.   A good estimate should be made beforehand of the total data expected and the required digitizing rate, to determine data storage requirements and computing load requirements;   otherwise both may quickly reach the reign of the impossible.   At least during the exploratory stage computation should be interactive, with the researcher calling at will on a number of manipulative subroutines and data display routines to look for useful and interesting features in the data.   The question of spectrum interpretation is to some extent a creative process on the part of the interpreter: by extensive examination of various types of computed results for each known type of flow, patterns emerge and can be used for future guidance.

10   Spectrum Interpretation

General remarks

In their well written book, Jenkins and Watts say 'Our present computing facilities are greatly in excess of our ability to make sense of practical data'. Ten years have passed, during which time computers have become even more powerful, while our ability to interpret spectra does not seem to have advanced much. In writing this chapter, we are embarrassed by not having much to say. However, in view of the importance of this aspect of spectral analysis, we would rather risk showing our ignorance, if only to start the reader thinking about the topic.

The simplest possible analysis one can perform on a computed spectrum is to compare it with a standard spectrum (template), familiar to anyone who has studied chemistry. When we wish to know if some substance is present in a given sample, we measure the radiation spectrum of the sample over a frequency range in which the spectrum of the substance is known, and compare the two. When the given substance is known to be a mixture, its spectrum would be some combination of the spectra of its constituents. But even then we may be able to find a frequency range over which the radiation emitted by the substance to be identified is dominant. We discuss the more difficult situation below.

Spectrum analysis allows us to identify sharp peaks. For example, in wind tunnel tests of aeroplane models, peaks in the vibration spectra of various parts of the model indicate the presence of resonances, which could lead to structural failure, and hence must be eliminated by design modifications. Similarly, the spectrum of road and airport runway surfaces after trend removal, should not contain sharp peaks, which indicate periodic structures that can cause severe vibrations in the vehicles travelling over them. Peaks in the spectrum of earthquake waves give information about the vibration modes, and hence the structure, of the earth. For electric signals, sharp peaks in spectrum nearly always show feedback paths in the producing system. Thus, if the input signal is a sinusoid of frequency f, and the system is such that every peak/trough in the input is reinforced by the feedback from the previous peak/trough, then the oscillation is greatly magnified. Similar effects are caused by negative feedback if the feedback from every peak goes to a trough, making it deeper, and that from each trough goes to a peak. There are many other interpretations of sharp peaks in a spectrum depending on the signal source. Of course, in all cases we must be sure that the peaks are genuine and are not caused by statistical fluctuations.

In many applications we can relate the frequencies to physical effects. We saw this in Chapter 9 where we discussed turbulence. Audial signal is another example, where frequency corresponds to pitch or tone. In other cases, one has to use imagination.

Composition analysis

Let us consider the determination of the composition of a mixture by analysis of its spectrum. The assumption here is that the contribution of each substance to the total radiation is of comparable importance and no constituent dominates. When each constituent spectrum is made up of sharp peaks (a line spectrum), it should be possible to separate out the various substances by careful examination. But when the peaks are mostly broad ones, the job becomes extremely difficult. We now present a technique for determining the composition by least-squares fitting of the mixture spectrum as a sum of constituent spectra.

Assume that the given sample contains constituents $1, 2, ..., m$, which have spectra $s^1, s^2, ..., s^m$. Each spectrum contains M values. Then the spectrum of the mixture is assumed to be

$$S_i = \sum_{j=1}^{m} s_i^j \, \sigma_j, \qquad i = 1, 2, ..., M, \tag{1}$$

where σ_j represents the percentage content of substance j (σ_j may be 0 but not negative). As a pre-condition the signals coming from the constituents are supposed to be independent and non-interfering. Otherwise equation (1) is not valid. We are thus assuming, for example, that radiation from material 1 is not absorbed by 2, nor does it cause induced emission from the others.

The procedure for computing the values of σ has already been outlined in Chapter 2. We try to minimize the mean square error

$$\sum_i \left(S_i - \sum_j s_i^j \, \sigma_j \right)^2, \tag{2}$$

with respect to each σ. This gives the equations

$$\sum_{k=1}^{m} \sigma_k A_{jk} = \sum_{i=1}^{M} s_i^j S_i, \qquad \text{for } j = 1, 2, ..., m,$$

with

$$A_{jk} = \sum_{i=1}^{M} s_i^j s_i^k . \tag{3}$$

These are analogous to equations (2) and (3) in Chapter 2, but of course A_{jk} is not diagonal now as the standard spectra s are not orthogonal with each other. Consequently (3) has to be solved as m linear equations in m unknowns, as distinct from equation (5) in Chapter 2.

Although the above procedure for finding σ is simple in theory, there is much potential hazard. First, the sample may contain substances not included among the assumed m constituents. Or, we might have assumed the spectrum of some chemical which is not actually present. Both conditions can produce significant errors in the computed values of σ, especially those of small magnitudes. To guard against such possibilities, one should always examine the residuals

$$S_i - \sum_j s_i^j \sigma_j . \tag{4}$$

If some of the values are quite large, then it is highly likely that the set of assumed constituents do not match the set actually present. Generally speaking, something wrongly left in causes smaller errors than something wrongly left out, but it is much harder to detect.

Sometimes it is useful to apply weighted least-squares fitting, i.e., each point in the spectrum is given a weight w_i, and we try to minimize $\sum_i w_i(S_i - \sum_j s_i^j \sigma_j)^2$, which turns (3) into

$$\sum_k \sigma_k \sum_i w_i s_i^j s_i^k = \sum_i w_i s_i^j S_i , \qquad \text{for } j = 1, 2, ..., m . \tag{5}$$

When we find large residuals but do not know what is causing them, it is worth while to define a set of weights inversely related to the magnitude of the residual at each point and try a weighted least-squares fit. By giving small weights to the bad points, we may obtain a better set of σ values. This is a robust estimation method. It is particularly effective when the wrongly left out (or wrongly left in) spectrum contains a few peaks.

Another potential problem is that of linear dependence, meaning that some of the standard spectra can be expressed as linear combinations of others. For example, if $s^1 = as^2 + bs^3$, then if the overall spectrum is equal to s^2, it could just as well be $(s^1/a - bs^3/a)$. The set of equations (3) has no unique solution. (Another way of saying this is that matrix A is singular and has no inverse.) Of course, few spectra are exactly related this way. Rather, it could be that some of the spectra are approximately equal to the linear combinations of others. In such a case (3) is said to be ill-conditioned. It can be solved, but the solution is quite unreliable. This unreliability is manifested in two ways: by cancellation and zero division. During the solution of the equations, we find ourselves subtracting numbers which are approximately equal, producing intermediate results which have very large relative errors. We sometimes also find ourselves dividing numbers by values of extremely small mangitudes, producing enormous and unreliable quotients.

Ill-conditioning is seldom a problem if m is small; but when m goes above 10 it is almost bound to occur. The problem has been extensively studied by numerical analysts. The treatment consists of the following three parts:

1. Use of double precision arithmetic in the solution of the equations;

2. application of pivoting; or

3. use of orthogonalization procedure.

These are discussed in recent books on numerical analysis. We provide some notes on orthogonalization in Appendix 1, Section 4.

The composition analysis technique seems to have most use for chemical substances, and a whole book (Blackburn) has been written on the subject. The method is applicable to any other signal that contains contribution from several sources. However, to use it we must know what the sources are, the spectrum of each, and that the signals from different sources are independent. These conditions are seldom met. In the next section we present a more common example.

System identification

The rather grand title of this section means in its simplest form: given input $x(t)$ and output $y(t)$, find h such that

$$y(t) = \int h(s)\, x(t-s)\, \mathrm{d}s + n(t),\tag{6}$$

where $n(t)$ is some random interference independent of x. We must of course have good reasons to believe that this is a fair approximation to the physical relation between x and y, which requires the system producing y to be linear and time invariant. (See Appendix 2, Section 1 for explanatory notes.) The problem looks easy, but this is deceptive. The reliability of computed values of h is usually extremely poor, and we can hope for little more than a rough idea about the system.

To derive the required results, we multiply both sides of (6) by $x(t+t')$ and take ensemble averages, giving

$$\langle y(t)\, x(t+t')\rangle = \langle x(t)\, y(t-t')\rangle = g(-t') = \int h(s)\, a(t'+s)\, \mathrm{d}s + \langle x(t+t')\, n(t)\rangle.$$

The last term is zero as n and x are assumed to be independent. Thus, g is produced by convolving h with a. It then follows that

$$G_i^* = H_i^*\, S_i^x.\tag{7}$$

Then we multiply (6) by $y(t+t')$ and take averages to give

$$\langle y(t)\, y(t+t')\rangle = \int h(s)\, g(t'+s)\, \mathrm{d}s + n(t)\, n(t+t')\rangle,$$

the last term again due to the independence between x and n. The above leads to

$$S_i^y = H_i^* G_i + S_i^n. \tag{8}$$

Comparison of (7) and (8) shows that $H_i^* = G_i^*/S_i^x = (S_i^y - S_i^n)/G_i$, or

$$S_i^n = S_i^y - |G_i|^2/S_i^y = S_i^y(1 - \kappa_i). \tag{9}$$

It is clear that if $n(t) = 0$ identically, then we must have $(1 - \kappa_i) = 0$. In other words, when y is linearly related to x and completely determined by it then the coherency spectrum is always 1. It is also clear that $\kappa_i \sim 0$ implies $S_i^y \sim S_i^n$.

Thus, in theory we can find h by computing $H_i = G_i/S_i^x$ followed by Fourier transformation. In practice the results thus obtained are extremely doubtful. As we said several times, even if the computed H is fairly close to what it should be, this does not guarantee that h is reliable too. In any case, H is not necessarily accurate. The trouble occurs because correlation between h and a gives g, but the estimated values of $a(t)$ and $g(t)$ must be modified by a window before S and G are calculated. Thus, the computed h is at best something that gives $w(t)\,g(t)$ when correlated with $w(t)\,a(t)$. It is sensitive to our choice of w. Also we must always inspect S_i^n, given by (9), to satisfy ourselves that it appears to make sense. If we have reasons to believe that n is white, while (9) gives a lopsided spectrum, then something is wrong. Perhaps x and y are not linearly related, or we used the wrong computing parameters.

11 Digital Filtering

General comments

Filters separate different parts of a mixture. A filter in signal processing separates different parts of a composite signal, e.g., low frequency versus high, good signal versus bad noise, etc. Many filters and much mathematical theory have been proposed on this subject, and we have such things as 'optimal smoothing/predicting filters' and 'Chebyshev band-pass filters'. Considerable effort has also been devoted to finding ingenious ways of improving available filters to approach some theoretical ideals.

We shall, however, take a very elementary look at this subject. In ordinary applications only relatively simple filters are needed. Further, some of the ideal filters people try to use are of dubious value, besides being quite difficult to implement. For example, the ideal low-pass filter is one that passes all frequency components below some limit unaltered while suppressing all other frequencies. But we saw on page 11 that such an operation will produce a result that shows large ripples near any sharp changes in the input signal (Gibb's phenomenon). Thus, even if we could implement the ideal low-pass filter cheaply it would be harmful to actually use it.

As mentioned in Chapter 4, one main use of digital filters in spectral analysis is to separate different parts of the input signal for different analyses. Another is to remove very prominent peaks from the spectrum to make it smoother and less prone to leakage. Because our filters are not ideal, we do not get complete rejection of unwanted frequencies; nor are the wanted frequencies always passed without change. However, the second is no problem at all, since we can always compute the correct spectrum of the wanted signal by making appropriate changes to compensate for the effect of the filters.

There is one point to note in filter design. Our discussion of windowing has shown us how to change the Fourier transform of a function by an appreciable amount but cause only minor changes in the reconstituted function. (The opposite can also happen: minor changes in one domain can make great differences in the other.) Consequently, in filter design it would be most unwise to judge the quality of a filter by how it looks in the frequency domain alone. It is always necessary to examine its effect in time space as well.

Filtering can be performed in either time or frequency space. In time space, we can use either recursive or non-recursive filtering. In the latter, the output at any time depends only on values of the input; in the former, it depends on previous output values as well as input values. Thus we have

$$y_i + \sum_{k=1}^{\infty} h_k^1 y_{i-k} = \sum_{k=-\infty}^{\infty} h_k^2 x_{i-k} . \tag{1}$$

Note that y_i can only depend on previous values of y, but may depend on *all values* of x.* However, in practice the two summations include only a small number of terms each. When $h_k^1 = 0$ for every k, then the filter is non-recursive; otherwise it is recursive.

If we put $h_0^1 = 1$, then we have

$$\sum_{k=0}^{\infty} h_k^1 y_{i-k} = \sum_{k=-\infty}^{\infty} h_k^2 x_{i-k} .$$

Both summations can be recognized as discrete convolutions. (Except they are in time space.) Thus, we define the Fourier series, which depend on f rather than t: $H^1(f) = \Sigma h_k^1 \exp(ckf)$, $Y(f) = \Sigma y_k \exp(ckf)$, etc. We have $H^1(f) Y(f) = H^2(f) X(f)$ or

$$Y(f) = X(f) H^2(f)/H^1(f) . \tag{2}$$

For non-recursive filtering, $h_k^1 = 0$ except for $h_0^1 = 1$. As result, $H^1(f) = \exp(0) = 1$, and so

$$Y(f) = X(f) H^2(f) . \tag{3}$$

The recovery of y from $Y(f)$ is the same as computing the Fourier coefficient, again with time and frequency spaces reversed:

$$y_i = \int_0^1 \exp(-cif) Y(f) \, df .$$

It is clear that the frequency content of y is very dependent on the expression $H^2(f)/H^1(f)$, which is given the name of *transfer function* and denoted as $H(f)$. It corresponds to quantities h_i in time space

$$h_i = \int_0^1 \exp(-cif) H(f) \, df .$$

Since Y is the product of H and X, y is related to h and x as a discrete convolution, so that

$$y_i = \Sigma h_k x_{i-k} . \tag{4}$$

* This is analogous to what statisticians call 'autoregressive-moving average processes'.

If the filter is non-recursive h and h^2 are obviously the same; the relation between h, h^1 and h^2 for a recursive filter is not easy to find without going to frequency space; h is called the time filter. If $h_i = 0$ except for m consecutive values of i, then it is said to have filter length m.

Obviously, X, Y, H etc., are closely related to the DFT of x, y, h etc., as

$$\hat{X}_i = N^{-1} \sum_{j=0}^{N-1} \exp(-cij/N)x_j = X^*(f)/N \quad \text{for } f = i/N,$$

or $X(f)$ is just $N\hat{X}_{Nf}^*$. If we know the spectrum of y and the transfer function we can easily find the spectrum of x, as

$$S_i^x = \langle |Y(i/N)/H(i/N)|^2 \rangle = S_i^y / |H(i/N)|^2, \quad \text{for } i = 0, 1, ..., \tfrac{1}{2}N.$$

As we are only interested in these frequencies f lies only between 0 and ½. If the N values are measured over $[0, L]$, then f corresponds to the physical frequency Nf/L, or f/Δ where Δ is the sampling interval L/N.

In digital filtering we can again just work with the standard intervals, and identify results with physical time and frequency values later.

Since $x(t)$ is real, we know $X(f) = X^*(-f)$. As long as h is real, or, as long as $H(f) = H^*(-f)$, we would have $Y(f) = Y^*(-f)$, so that y is real. If, however, h is symmetric as well, $h_{-i} = h_i$, we would also have $H(f) = H(-f) = $ real. This would mean that there is no change to the phases of the components of x. There are only changes in the magnitude of each X_i by its multiplication into $H(i/N)$.

Our job in filter design is then to find H and h with prescribed shape by suitable choice of h^2 and, perhaps, also h^1. Generally speaking, a recursive filter with the same number of parameters as a non-recursive filter will have a better performance. The reason lies in the greater accuracy with which we can approximate a given function by a rational function $H^2(f)/H^1(f)$, than by a polynomial, $H^2(f)$ alone. (This is well known to numerical analysts.) Another way of saying the same thing is that 'feedback improves performance at little increase in complexity' — well known to control engineers. But recursive filters do have the problem that $H(f)$ is not purely real because h^1 is not symmetric, since it is zero for negative values of k and non-zero for some positive values of k. Consequently, multiplication of $H(f)$ into $X(f)$ causes phase changes, which may not always be desirable. In comparison $H^2(f)$ can always be made real by setting $h_k^2 = h_{-k}^2$. Non-recursive filters are also easier to design. Finally recursive filters may be unstable, i.e., its output may increase without limit or oscillate with increasing amplitude in response to stable input. This is because $H^1(f)$ may be 0 for some f so that $H(f) \to \infty$. Non-recursive filters, with $H^1(f) = 1$, do not have such problems.

Because filtering according to (4) is equivalent to multiplying $X(f)$ by $H(f)$ to produce $Y(f)$, it is natural to try the following: Fourier transform x to obtain X_i, modify each coefficient by some value H_i to produce Y_i and then inverse transform to produce y. There is, however, a significant difference between the two filter implementations. Convolution (4) is non-cyclic, while $H_i X_i$ produces the cyclic convolution between x and h. In practical terms it amounts to this: in (4) the first few values of y depend on non-existent values of x, while in a cyclic convolution they would depend on the first few values of x as well as the last few. Neither is nice to have. As a result, the beginning and ending portions of filter output are unreliable and should usually be discarded. Further comments on this situation are given in the next section.

When the input signal is such that its 'head' is very appreciatively different from its 'tail', e.g., when it increases monotonically, then cyclic convolution, mixing the head and the tail to produce the output, would produce very bad results. Consequently, trend removal to make the whole input reasonably uniform in behaviour throughout the interval helps a lot. This fact is much less pronounced for non-cyclic convolution.

When performing filtering by multiplication in frequency space, it is tempting to use a set of H_i that have 'the desired shape'. For example, if we believe x contains low frequency signal and high frequency noise, we might be tempted to simply reduce the high index X values to zero. As we said before, this is bad practice because we must consider the operation in both frequency and time spaces. Suppressing the high frequencies like this will give rise to Gibb's phenomenon, since in time space we are convolving the input with a highly oscillatory function. Generally speaking, we should never use any H that abruptly cuts off certain frequency ranges. Instead, we must gradually terminate the non-zero portions of H, as in windowing, to ensure that h varies smoothly in the time domain.

Finally, when we are filtering a long vector, say one having N elements, using an h of filter length M, it is possible to slice up the vector into M element blocks and achieve the filtering using $2M$-point FFT, in an arrangement similar to the computation of correlation as explained in Chapter 6. We leave it to the reader to figure out the details.

Should data be tapered?

Before discussing actual filters, let us first study the question of tapering in digital filtering, something we have already encountered earlier under the name of linear windowing. It has been suggested that we should multiply the data by a linear window that gradually terminates the two ends of the vector before we take its Fourier transform and apply the filter transfer function H. Otherwise, it is argued, because of leakage frequency components which are supposed to be removed by the filter will get mixed with those supposed to be kept, reducing the effectiveness of the filter. This point is said to be particularly significant when N is not a power of 2, in which case we could add zeros to the data vector in order to use FFT. This produces an abrupt change within the data vector and hence leakage.

The fault in the argument is this: a 'good' Fourier transform does not guarantee a good time function. We have seen this again and again.

A truncated Fourier series has a more accurate Fourier transform than the Fejer series, but is not as well behaved in the time domain. In windowing, we improve the power spectrum by drastically altering the autocorrelation. Whereas Ξ^w may be a good approximation to S, its corresponding time function, $\xi_a(t)\,w(t)$, certainly is not a good approximation to $a(t)$. Thus, while tapering might produce a smooth Fourier transform, this is no guarantee that we will get a good y.

Let us consider the whole operation in the time domain. The data vector after tapering is $w_i\,x_i$, so that the output of the filter is now

$$y_i = \sum_j h_j\,w_{i-j}\,x_{i-j}\,.$$

The effect of the w values is to reduce the magnitude of, say, the first and last m values of x, regardless of what x and h are. Suppose h is of length M, and $M < m$, then only y_i with $i \sim 0$ and $i \sim N$ are affected by the tapering. For i away from the ends y_i depends only on those values of x that multiply into $w = 1$. But for i near the ends, y_i is the function of a set of numbers $w_{i-j}\,x_{i-j}$, which are completely different from the input data x_{i-j}. How that is going to produce better results is beyond our understanding. As far as we can see, tapering in this case has either no effect, or where it does, has an adverse effect. By tapering, we have effectively thrown away the beginning and end portions of our data, for the unclear benefit of making the two ends vary smoothly with time.

The benefit of tapering appears more convincing when M is large, in which case y_i for $i \sim 0$ and $i \sim N$ includes non-existent x that have indices outside the range $[0, N]$. Another way of saying the same thing is that the filter takes time to settle, i.e., y_i is not reliable until i exceeds M so that y_i no longer depend on undefined values of x. By tapering the two ends and eliminating the abrupt changes, we obtain less oscillatory output values. However, we can achieve the same purpose without throwing away valuable data. Instead of completely altering the first and last m values of the input, we could add at each end a set of artificial input values and taper *them*, while leaving the real data unchanged or only slightly changed. The artificial data should resemble the real data. For example, if the data are fairly smooth, we could simply extend them by extrapolating the two ends as two straight lines. If the given data show oscillation superimposed on a smooth change, we would add a similar oscillation to the extrapolated values. The extrapolations can then be tapered. Alternatively, we can also taper some of the data along with the extrapolations. As the tapering window goes gradually from 1 to 0, it will change the data only slightly.

Clearly, there can be no such thing as a universally applicable tapering window. The values of m, which decide the amount of data that are altered, depend on the settling time of the filter. The above again illustrates the importance of time domain considerations in any filter design. Generally speaking, tapering is of the greatest benefit when used with recursive filters, which depend effectively on the infinite past of x, and hence have long settling times. When filtering is done by Fourier transformation, trend removal is more important than tapering. In particular, when we perform low-pass filtering on data with a clear trend, it is essential that the trend be removed first and perhaps added back after filtering. (This may also be useful in recursive filtering, as such filters can be greatly affected by the trend.) It is also useful to note that, if trend removal produces a data vector that starts and ends with value 0, then tapering is probably unnecessary as the two ends are already quite gradual.

What we said about tapering applies to the actual filtering of x. When we are *designing* the filter it is indeed useful to taper x in order to get a better picture of $X(f)$. The whole business comes down to this: if the result one wants is a function of time, taper in frequency space (e.g., Fejer series); if it is the transform that we want, taper in time space. Tapering two transforms ahead is usually not helpful.

Non-recursive filters

Let us consider the following filter (m is an odd integer):

$$h_j = 1/m; \quad \text{for } j = -\tfrac{1}{2}(m-1), ..., \tfrac{1}{2}(m-1); \quad h_j = 0 \text{ otherwise}. \quad (5)$$

Clearly,

$$H(f) = m^{-1} \sum_j \exp(cjf) = \sin(m\pi t)/[m \sin(\pi t)],$$

derived in the same way as equation (21) in Chapter 2. Since $H(f) = 1$ at $f = 0$ and $f = 1$, but is only m^{-1} at $f = \tfrac{1}{2}$, this is a crude low-pass filter. Recall that multiplication between h and x gives cyclic convolution between H and X, so that H is periodically extended outside the interval $[0, 1]$. As a result, f is the same as $(f-1)$, or, the interval $[\tfrac{1}{2}, 1]$ is the same as $[-\tfrac{1}{2}, 0]$. The described filter passes with little change those $X(f)$ with $f \sim 0$, but attenuates those with $f \sim \pm\tfrac{1}{2}$. (In physical units these are just $f = \pm 1/2\Delta$, the maximum frequency measurable when sampling at intervals of Δ.)

Already, it is clear filter design is rather similar to windowing. (The very idea of filtering is to get a window through which x looks prettier.) The filter just discussed is not a good low-pass filter because $H(f)$ does not decrease rapidly as $f \sim \pm\tfrac{1}{2}$, and it contains negative side lobes. Techniques for improving things are just the same bag of tricks we used in windowing. Let us suppose that we use h on x twice,

$$z_i = \sum_j h_j y_{i-j} \qquad \text{and} \qquad y_j = \sum_k h_k x_{j-k}.$$

Obviously $Z(f) = [H(f)]^2 X(f)$, so that the overall effect is to apply the transfer function

$$\sin^2(m\pi f)/[m^2 \sin^2(\pi f)].$$

Compare this with equation (26) in Chapter 2, and note its analogy to the Fejer series. It is also related to the Bartlett window. The reason for the latter is simple: the filter producing z is just h convolved with itself (as h

is used twice), which produces

$$h'_j = (m - |j|)/m^2, \quad \text{for } j = m+1, ..., m-1; \quad h'_j = 0 \text{ otherwise}.$$

The connection with Fejer–Bartlett is obvious. The transfer function is 1 at $f = 0$ and m^{-2} at $f = \pm\frac{1}{2}$. Its main lobe extends from $f = -m^{-1}$ to $f = m^{-1}$. (In physical units $-1/(m\Delta)$ to $1/(m\Delta)$.) Outside the interval the values are quite small. Thus, this is a more effective low-pass filter.

Of all the low-pass filters the one just discussed is the most economical since its two-stage computation takes on $2Nm$ additions. A further economy comes in when we consider the following: y_{i+1} is x summed from $j = i+1-\mu$ to $i+1+\mu$, where $\mu = (m-1)/2$, while y_i is x summed from $j = i-\mu$ to $i+\mu$. Hence $y_{i+1} - y_i = m^{-1}(x_{i+1+\mu} - x_{i-\mu})$. In other words, y_{i+1} can be produced from y_i with only two additions. Hence the whole vector y takes only $2N$ additions to evaluate, and the same for vector z. The filtering operation requires only $4N$ additions to complete, which is virtually unbeatable. We could go further and convolve z again with h, but it quickly goes to the point of diminishing returns.

Now let us consider the band-pass filter. The low-pass filter has its main lobe centred on $f = 0$. Suppose there are two main lobes, centred on $f = \pm\phi$. This passes frequencies close to $\pm\phi$ but rejects the others. The function $H(f)$ must be symmetric, i.e., $H(f) = H(-f)$, to give a real and symmetric h. We get such a transfer function by convolving a low-pass transfer function with a function of f that has two sharp peaks at $f = \pm\phi$. This corresponds to multiplying h_j by the sum of two exponentials $\exp(c\phi j) + \exp(-c\phi j) = 2\cos(2\pi\phi j)$. (We remind the reader that $0 < \Phi < \frac{1}{2}$ and that physically ϕ corresponds to frequency ϕ/Δ.) Thus, a crude band-pass filter is given by

$$h_j = m^{-1}\alpha\cos(2\pi\phi j), \quad \text{for } j = -\mu, ..., \mu; \quad h_j = 0 \text{ otherwise}.$$

A better ones is the product of $\cos(2\pi\phi j)$ and (5), namely

$$h'_j = m^{-2}\alpha\cos(2\pi\phi j)(m - |j|), \quad \text{for } j = -m+1, ..., m-1.$$

($m = 2\mu+1$ for similar bandwidth.) The factor α needs an explanation: it is supposed to be 2 to ensure that the two peaks in $H(f)$ have height 1. This would mean that there is no amplification. However, when ϕ is too small, $\phi < 1/m$, the two peaks partly overlap, so that its maximum height may exceed 1 if we do not make α smaller. When ϕ is very close to $\frac{1}{2}$, the two peaks merge also. In particular, when $\phi = \frac{1}{2}$, the two peaks coincide, and we get one lobe centred on $f = \frac{1}{2}$, which is the same as $f = -\frac{1}{2}$ because of periodic extension. Or, we have half of the peak extending from $f = -\frac{1}{2}$ to some negative value, and the other half of it extending from some positive frequency to $f = \frac{1}{2}$. This then is a high-pass filter. By choosing α to be 1 we ensure that the height of the peak is 1. Thus, we get a high-pass filter by multiplying h of the low-pass filter by $\cos(2\pi\phi j) = (-1)^j$. So a crude high-pass filter is given by

$$h_j = (-1)^j/m, \quad j = -\mu, ..., \mu, \quad \text{for } m = 2\mu+1; \quad h_j = 0 \text{ otherwise}.$$

Again this can be applied to x twice to get a better filter, which has the combined effect of the following filter: $h'_j = m^{-2}(-1)^j(m-|j|)$, $j = -m+1, ..., m-1$.

The band-pass filter described above requires $2NM$ multiplications and additions to implement, and this cannot be simplified by the previous trick used on low-pass filters. However, the trick can be used on the high-pass filter, which is implemented using $4N$ signed additions. We leave it to the reader to figure out the details.

Another way of deriving band-pass or high-pass filters is to consider shifting $X(f)$ rather than $H(f)$. This is done by multiplying x_i by $\exp(\pm c\phi j)$. In real arithmetic the process is as follows: multiply x_j by $\cos(2\pi\phi j)$, apply any low-pass filter, multiply by $\cos(2\pi\phi j)$ again; do the same using $\sin(2\pi\phi j)$ instead of $\cos(2\pi\phi j)$, add the two outputs. The reader might try to analyse the process and see why it works.

Sometimes it may be a good idea to implement a band-pass filter by a combination of high- and low-pass filters. If we wish to pass only the frequencies $[\phi_1, \phi_2]$, we would first do a low-pass filtering up to the frequency ϕ_2, followed by a high-pass filtering that extracts the frequencies $[\phi_1, \frac{1}{2}]$. The whole operation can be implemented using $8N$ additions. Note that some amplification is required to ensure max $H(f) = 1$.

Recursive filters

We shall look at only one example of recursive filters, to show both their potentials and their shortcomings. The following is a *first-order* recursive filter, as the y_i varies with only one previous output:

$$y_i = \alpha y_{i-1} + (1-|\alpha|)x_i, \text{ i.e., } h^1_1 = \alpha, h^2_0 = 1-|\alpha|; \ h^1, h^2 = 0 \text{ otherwise.}$$

By using (2) we have

$$H(f) = (1-|\alpha|)/[1-\alpha\exp(cf)] . \tag{6}$$

This in general is not purely real. To see its filtering effect more clearly we compute

$$|H(f)|^2 = (1-|\alpha|)^2/[1+\alpha^2-2\alpha\cos(2\pi f)] . \tag{7}$$

If $|\alpha| \geqslant 1$, the denominator may be 0 for some f, so that the filter becomes unstable. If $1 > \alpha > 0$, then $|H(f)|^2 = 1$ at $f = 0$ but decreases as $f \to \pm\frac{1}{2}$. This gives a low-pass filter. But if $-1 < \alpha < 0$, then $|H(f)|^2 = 1$ at $f = \pm\frac{1}{2}$ and decreases towards $f = 0$. This is a high-pass filter. To find the band-width for the filter, we note that near $2\pi f \sim 0$ we have the approximation $\cos(\Theta) \sim 1-\Theta^2$. For $\alpha > 0$ this gives $|H(f)|^2 \sim [1+40\alpha f^2/(1-\alpha)^2]^{-1}$. Starting at 1 for $f = 0$, this drops to 0.5 at $\pm(1-\alpha)/(2\pi\sqrt{\alpha})$, 0.2 at $\pm(1-\alpha)/(\pi\sqrt{\alpha})$, and 1/17 at $\pm2(1-\alpha)/(\pi\sqrt{\alpha})$. Thus we could say roughly that the filter passes components between $f = \pm(1-\alpha)/3\sqrt{\alpha}$. For $\alpha < 0$ the corresponding frequencies are $\pm[\frac{1}{2}-(1-\alpha)/3\sqrt{\alpha}]$. Note that these formulae are accurate only if $|\alpha| > 2/3$. As α goes below $\frac{1}{2}$, $|H(f)|$ becomes rather flat, and can no longer even be called a low/high pass filter as it only attenuates some of the frequencies slightly. (See Fig. 11.1.)

The implementation of the above recursive filter is very simple indeed. Whereas narrow band non-recursive filters require very large m, here we merely adjust α appropriately. However, we must not make $|\alpha|$ too close to 1 as then the filter is close to being unstable and takes very long to settle, either at the start or after any sudden change in the input. We also note that recursive filters need much fine tuning, indicated by the fact that, at $\alpha = \frac{1}{2}$ and $\alpha = \frac{1}{3}$ the same filter behaves very differently. Even a small change in α can make a large difference in the results. Finally, implementing a band-pass filter by combination of low/high pass filters is unlikely to work here as the described recursive filter does not have a sharp cutoff.

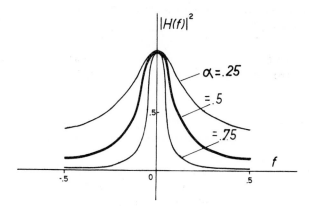

Fig. 11.1

By introducing more complicated forms of h^1 and h^2 we can obtain various other filters, but the mathematics is too involved for us to present here. It suffices to say that, we can write down the desired shape of $H(f)$ in terms of trigonometric functions or Chebyshev polynomials, and then convert these to expressions for h^1 and h^2. Details can be found in Chapter 3 of Otnes and Enochson and references quoted there, as well as in the numerous books on digital signal processing listed on page 152.

Two useful filters

We shall now look at two other filters of interest. The first, the optimal smoothing filter, is designed to reduce the average power of additive noise mixed with signal. The second, the differentiation filter, is for computing the derivative of a noisy function. The discussion will be rather brief.

Suppose we have the input signal $x(t) = u(t) + v(t)$, where $u(t)$ is 'good' signal and $v(t)$ is noise. We wish to produce $y(t)$ such that its mean-square deviation from $u(t)$ is as small as possible. To do this by filtering, we look at

$$y(t) = \Sigma \, \exp(cit) Y_i = \Sigma \, \exp(cit) H_i \, X_i \,,$$

and require H_i to be such that the following is minimized:

$$\int \langle [y(t) - u(t)]^2 \rangle \, dt = \langle y(t)^2 \rangle - 2\langle y(t) \, u(t) \rangle + \langle u(t)^2 \rangle \,.$$

Expressing everything as Fourier series we get

$$\sum_{j,k} \int \exp(cjt + ckt) \, dt \langle X_j \, X_k \rangle H_j \, H_k - 2 \sum_{j,k} \int \exp(cjt + ckt) \, dt \langle X_j \, U_k \rangle H_j + \int \langle u(t)^2 \rangle \, dt$$

$$= \sum_{j} |H_j|^2 \langle |X_j|^2 \rangle - 2 \sum_{j} H_j \langle X_j \, U_j^* \rangle + \int \langle u(t)^2 \rangle \, dt \,,$$

where we have used the relations $\int \exp(cjt + ckt) \, dt = \delta_{j-k}$, and $H_{-j} = H_j^*$. Differentiating the expression with respect to H_i and setting each partial derivative to zero give:

$$H_i = \langle X_i \, U_i^* \rangle / \langle |X_i|^2 \rangle \,. \tag{8}$$

Obviously, $X_i = U_i + V_i$. Usually we can make the assumption that the noise is independent of the signal, or $\langle V_i \, U_i^* \rangle = 0$. This leads to

$$H_i = S_i^u / (S_i^u + S_i^v) \,. \tag{9}$$

Thus, $H_i \sim 1$ if $S_i^u \gg S_i^v$, and $H \sim 0$ if $S_i^u \ll S_i^v$, so that components rich in signal are passed with little attenuation, while those rich in noise are reduced.

The optimal smoothing filter is only a crude device, but for the usual problems it is quite adequate. An important point to remember is that we must take care to avoid Gibb's phenomenon. H_i must be made to decrease gradually from frequencies where S^v is small to those where it is large. Consequently, the values given by (9) are to be used as rough guides rather than to be applied blindly. Some modifications to eliminate undesirable oscillations, at the expense of the mean-square error, may be necessary.

Turning our attention to differentiation, let us first consider the obvious method, that of finite difference: $y_i = (x_{i+1} - x_{i-1})/2\Delta$, where Δ is the sampling interval. The above is called a central difference. The problem is that often x is not exact. It may suffer from measurement errors, digitization errors or roundoff. If x has error between $\pm\delta$, y has error between $\pm 2\delta/2\Delta$. Since Δ is a small number, the error is greatly magnified.

We can also look at the situation in the following way: the above is a filtering operation with $h_{-1} = 1/2\Delta$ and $h_1 = -1/2\Delta$, while all other values of h are zero. This has the transfer function

$$H(f) = [\exp(-cf) - \exp(cf)]/2\Delta = \sqrt{-1} \, \sin(2\pi f)/\Delta \,.$$

Thus, $|H(f)|$ starts at 0 for $f = 0$, and increases as $f \to \pm\frac{1}{4}$. Now, x tends to vary much more slowly than the errors, so that $X(f)$ is dominated by low frequency while the noise spectrum has much high-frequency content.

By giving little weight to low-frequency components $H(f)$ causes magnification in the relative error. This is a problem common to all differentiating algorithms, since in every case we are trying to approximate dx/dt, which is

$$\frac{d}{dt} \, \Sigma \, X_i \exp(cit) \qquad \text{or} \qquad c \, \Sigma \, iX_i \exp(cit)$$

The presence of i inevitably magnifies the relative importance of noise.

It is clear that we can improve the derivative by filtering. We can reasonably assume that the noise spectrum in the input data is approximately flat (which is not strictly true as, for example, roundoff errors do depend on the numbers being rounded, but the relation is complex and the errors end up looking random), while the 'signal' varies slowly, so that some kind of low-pass filtering on y will produce good results. The only problem is to find the H. An earlier suggestion (Anderssen and Bloomfield, *Technometrics*, 1974, **16**, 69-75) is that we compute the power spectrum of x in the usual way, divide it into two parts: a constant background taken to the noise spectrum S^v, and the remainder as signal spectrum S^u, and finally apply the optimal smoothing filter (9) to the central differences. The method, though much faster and more accurate than earlier methods, still requires much computing time. If one is not too demanding about accuracy the following simple method might be tried. First smooth the central differences by the 'Tukey window' $y_i^1 = \frac{1}{4}y_{i-1} + \frac{1}{2}y_i + y_{i+1}$. If the result is still quite erratic, do the same to y^1 to product y^2, or even yet again to produce y^3. The $y^{1,2,3}$ are respectively y subject to low-pass filtering with filter lengths $m = 3, 5$ and 9. Longer filters might be used if it is believed that the true derivative changes only slightly over the length of the filter, but such is not likely to occur very often. When we do use a long filter the more economical methods outlined on page 121 should be applied, rather than a repeated use of the Tukey window.

12 Generalized Spectral Analysis

Optimal transform

The optimal transform, also called Kahunen–Loeve transform, is an orthogonal transform that is 'best' in a mathematical sense for representing a random process with a known autocorrelation function. The technique aroused much interest among information processing people, particularly those engaged in pattern recognition, during the past decade or so. Its experimental performance, however, did not live up to the earlier expectations. (Whose does? one may ask.) We are accordingly sceptical about its value in spectral analysis too. Still, it is worth while to take a brief look at it, if only to gain some flavour of the mathematics, which is used in numerous applications and seems inescapable wherever one turns. Some notes on the mathematics are provided in Appendix 1.

Consider a random process $x(t)$ with autocorrelation function $a(s, t)$ and a set of orthogonal functions $y_i(t)$. How do we measure the performance of the set for the purpose of representing $x(t)$? Once again let us try the mean square error: $E = \langle \int |x(t) - \sum X_i y_i(t)|^2 \, dt \rangle$. Remember $|A|^2 = A^*A$, which gives the relation

$$E = \int \langle x(t)^2 \rangle \, dt + \sum_{i,j} \langle X_i^* X_j \rangle \int y_i^*(t) y_j(t) \, dt - 2 \sum_i \langle X_i^* \int y_i^*(t) x(t) \, dt \rangle.$$

Because of orthogonality, the second term is $\sum \langle X_i^* X_i \rangle$. The last term is $-2 \sum \langle X_i^* X_i \rangle$. Thus

$$E = a(0) - \sum_i \langle |X_i|^2 \rangle,$$

where we have made use of the orthogonality of y. The above is minimized if we choose y values that maximize the second term. It turns out that the maximizing set satisfies the following relation

$$\int a(s, t) y_i(t) \, dt = \lambda_i y_i(s). \tag{1}$$

Before proving that such a set does minimize E, we need a little mathematical jargon. When we integrate $a(s, t)$ with any function $f(t)$, we produce a new function $g(s)$. This gives rise to the term *operator*. The expression $\int a(s, t)\, dt$ operates on $f(t)$ and turns it into $g(s)$. Now $y_i(t)$ has the particular property that operating on it with $\int a(s, t)\, dt$ gives rise to $y_i(s)$ times a constant λ_i. We call $y_i(t)$ an *eigenfunction* of the operator, and λ_i an eigenvalue. The significance of (1) becomes clearer when we note

$$\langle X_i\, X_j^* \rangle = \langle \int y_i^*(s)\, x(s)\, ds \int y_i(t)\, x(t)\, dt \rangle = \int y_j^*(s)\, a(s, t)\, y_i(t)\, dt\, ds$$

$$= \lambda_i \int y_j^*(s)\, y_i(s)\, ds = \lambda_i\, \delta_{ij}, \tag{2}$$

and $\langle |X_i|^2 \rangle = \lambda_i$.

Equations (2) show that the X values are uncorrelated. Our claim is then that uncorrelated orthogonal coefficients are also the 'best' in the sense of minimizing E.

To prove this claim, let us take another set of orthogonal functions $z_j(t)$. We write

$$X_j = \int x(t)\, z_j(t)\, dt \qquad \text{and} \qquad Z_i^j = \int z_j(t)\, y_i^*(t)\, dt. \tag{3}$$

Since y and z are both orthonormal and complete sets, we can expand $z(t)$ in terms of $y(t)$, and vice versa, using the coefficients (3). We can consider Z_i^j as the jth coefficient of $y_i^*(t)$ for the z-series, or as the ith coefficient of $z_j(t)$ for the y-series. Hence

$$z_j(t) = \sum_i Z_i^j\, y_i(t); \quad y_i(t) = \sum_j Z_i^{j*}\, z_j(t). \tag{4}$$

And we have

$$\delta_{jk} = \int z_j^*(t)\, z_k(t)\, dt = \sum_{ii'} Z_i^{j*}\, Z_{i'}^k \int y_i^*(t)\, y_{i'}(t)\, dt = \sum_i Z_i^{j*}\, Z_i^k.$$

Similarly, $\int y_i(t)\, y_i^*(t)\, dt = \sum_j Z_i^{k*}\, Z_{i'}^k = \delta_{ii'}$.

In particular, if we take only a finite set of M functions, we would have

$$\sum_j |Z_i^j|^2 \leqslant 1.$$

(This expression equals unity if the summation includes a complete set of Z values.)

Let us assume that $\lambda_1 \geqslant \lambda_2 \geqslant \lambda_3 \geqslant \ldots$. (We can always ensure this by rearranging the ordering of the y values if necessary.) If we take only the first M value of y, then

$$E = a(0) - \sum_{i=1}^{M} \lambda_i, \tag{5}$$

while for M values of z we have

$$|X_j|^2 = \int z_j^*(s)\, a(s, t)\, z_j(t)\, ds\, dt = \sum_{ii'} Z_{i'}^{j*}\, Z_i^j\, y_i^*(s)\, a(s, t)\, y_{i'}(s)\, dt\, ds$$

$$= \sum_{ii'} Z_{i'}^{j*}\, Z_i^j\, y_i^*(s)\, \lambda_i\, y_i(s)\, ds = \sum_i |Z_i^j|^2\, \lambda_i,$$

so that for the set of z values

$$E = a(0) - \sum_i \lambda_i \zeta_i ,\tag{6}$$

where $\zeta_i = \sum_{j=1}^{M} |Z_i^j|^2 \leqslant 1$,

If, however, we add up a complete set of values of ζ then the sum is equal to M, as

$$\sum_i \zeta_i = \sum_{j=1}^{M} \sum_i |Z_i^j|^2 = M.$$

Both $\sum_{i=1}^{M} \lambda_i$ and $\sum_i \zeta_i \lambda_i$ are weighted sums of λ values, and the total weight in each case is M. However, whereas the first sum included the largest M values of λ only, the second includes other, smaller terms while giving less weight to the largest values of λ as $\zeta_i \leqslant 1$. Consequently

$$\sum_{i=1}^{M} \lambda_i \geqslant \sum_i \zeta_i \lambda_i .$$

Hence, the z values cannot give a smaller mean-square error than the y values. In the mean-square sense, the performance of the y values is as good as we can hope for.

The above discussion is mathematically interesting, involving as it does the familiar concept of eigenfunctions and eigenvalues. It appeals to engineers and statisticians, for somewhat different reasons. To the former, the optimal transform extracts the maximum signal power from $x(t)$; the latter see merit in the lack of correlation between the X values, so that each is an independent piece of information. All these are very nice, but they do not guarantee us what we want for spectral analysis; namely y values that are well matched to the physical structures present in $x(t)$. In some problems eigenfunctions of the autocorrelation are physically meaningful; in most, physical interpretations are rather tenuous. Finally there is the problem of cost: the optimal values of y are not easy to find, and even after they have been specified it is still costly to apply the optimal transform on x.

For the optimal transform the power spectrum is just the set of eigenvalues λ_i, as λ_i is equal to $\langle |X_i|^2 \rangle$. However, we cannot compute X_i until we know y_i, which is defined by (1). But as (1) defines λ_i at the same time, it is not really necessary to find X_i at all for computing λ_i. It might be thought that we should, when given measured values of x, estimate ξ_a and then find its eigenvalues as the power spectrum. This is unfortunately of no use whatever, because if we make another set of measurements and compute a new ξ_a, we would obtain a spectrum defined in terms of a different set of y values. Since the two power spectra are not defined with the same orthogonal transform, we cannot even compare them meaningfully. A more useful approach is the following. Assume a theoretical autocorrelation, find y values from this according to (1), and then compute λ from measured values of x by averaging $|X_i|^2$ or by the use of the following relation

$$\lambda_i = \int y_i^*(s)\, \xi_a(s,\, t)\, y_i(t)\, dt\, ds\,, \tag{7}$$

which is just another way of writing $\langle |X_i|^2 \rangle$. This approach is theoretically feasible but unsatisfactory, since the assumption of an $a(\ )$ is somewhat arbitrary. All in all, we are doubtful of the value of optimal transforms in spectral analysis.

Fourier transform as optimal transform

Given an $a(s,\, t)$, it is possible to find the optimal transform. Now given an orthogonal transform, can we find an $a(s,\, t)$ for which it is optimal? This turns out to be quite easy. Conjugation of both sides of (1) gives $\int a(s,\, t)\, y_i^*(t)\, dt = \lambda_i\, y_i^*(s)$. (Note that a and λ are both real. Remember $\lambda_i = \langle |X_i|^2 \rangle$, obviously real.) If we consider $a(s,\, t)$ as a function of t, then the above relation says that its ith orthogonal coefficient is equal to $\lambda_i\, y_i^*(s)$. Consequently, $a(s,\, t)$ can be expressed as the following series

$$a(s,\, t) = \sum_i \lambda_i\, y_i^*(s)\, y_i(t)\,. \tag{8}$$

For the case of Fourier transform we have

$$a(s,\, t) = \sum_i \lambda_i \exp[ci(t-s)]\,. \tag{9}$$

Now, since $a(s,\, t)$ is supposed to be the autocorrelation of some random process, presumably time invariant, $a(s,\, t)$ must satisfy the following

$$a(s,\, t) = a(t,\, s) = a(|t-s|)\,,$$

which is assured if we have $\lambda_i = \lambda_{-i}$. However, at the same time (9) shows that

$$a(s,\, t) = a(|s-t|-1) = a(1-|s-t|)\,, \tag{10}$$

because $\exp[ci(t\pm 1)] = \exp(cit)$. In particular we must have

$$a(0) = a(1) = a(-1)\,.$$

This condition is not satisfied by many random processes, if any. As a result, the Fourier transform is not optimal for ordinary signals.

However, from page 61 we have the following expression for Fourier coefficients

$$\langle X_i^* X_j \rangle = \frac{1}{\pi(i-j)} \int_0^1 a(t)[\sin(2\pi jt) - \sin(2\pi it)]\, dt; \quad \text{for } i-j = \text{a non-zero integer.}$$

The expression vanishes if $a(t)$ is even over $[0, 1]$. (In other words, when (10) is satisfied.) But even when the condition does not hold the expression still gives small values. Thus, these Fourier coefficients are approximately uncorrelated, and could be said to be close to being optimal. But it should be remembered that, to recover a from S we require both integer and non-integer frequency terms. In contrast, (8) shows that for an optimal transform the uncorrelated terms alone would specify $a( )$ completely. In short, for optimal transform the statistically independent terms are also arithmetically independent, and vice versa, which does not hold for Fourier transform of time invariant random processes over a finite interval.

Walsh transform

In the past few years a long-standing but obscure set of orthogonal functions aroused much attention. They are the Walsh functions, proposed in 1923 by Joseph L. Walsh of Harvard University Mathematics Department. (That was before Walsh made important contributions to approximation theory, in particular theory of splines.) We shall give a short exposition of spectral analysis using the Walsh transform.

There are several forms of Walsh functions, of which we shall select one due to R.E.A.C. Paley (who co-authored a book on the Fourier transform with Norbert Wiener). We define

$$pal(i, t) = (-1)^{\Sigma\, i_k t_k} ,$$

where i_k, t_k are the binary digits of i and t ($k = 1, 2, ...$),

$$i = \Sigma\, 2^{k-1} i_k \qquad \text{and} \qquad t = \Sigma\, 2^{-k} t_k .$$

It is obvious that $pal(i, t) = \pm 1$. Further, we have the property

$$pal(i, t)\, pal(j, t) = pal(i \oplus j, t) \qquad \text{and} \qquad pal(i, t)\, pal(i, s) = pal(i, t \oplus s) ,$$

where $\oplus$ is XOR (exclusive OR, alias addition modulo 2). These two relations are readily derived from the definition of pal. The first 16 Walsh-Paley functions are shown in Fig. 12.1.

As before, we define

$$S_i = \langle X_i^2 \rangle = \int pal(i, t)\, pal(i, s)\, a(t-s)\, dt\, ds ,$$

without the absolute value sign, since pal and hence X are real. We have

$$S_i = \int_{-1}^{1} R(i; t)\, a(t)\, dt , \qquad (11)$$

where

$$R(i; t) = \int_{0}^{1-|t|} pal(i, s)\, pal(i, s + |t|)\, ds .$$

Whereas the Fourier power spectrum has a simple relation with a, the Walsh power spectrum does not. The structure of $R(i; t)$ is quite complicated and cannot be expressed in an easily evaluated form. Computationally, however, the Walsh power spectrum is extremely easy to obtain. There are, as before, several different ways to proceed, via autocorrelation or directly, by segment averaging or moving averages. The notable point is, however, that it is now possible to obtain completely identical results from different methods of averaging provided we choose $T = 2^{-m}$ for some integer m. To be specific, we have

$$\int_{-2^{-m}}^{2^{-m}} R(i/2^m, 2^m\, t)\, \xi_a(t)\, \mathrm{d}t = \sum_{j=i}^{i+2^m} X_j^2, \quad \text{for } i = 0, 2^m, ..., 2^n - 2^m.$$

The left-hand side is the ith indirect spectral estimate with a rectangular window of cutoff 2^{-m}, while the right-hand side is the ith moving average over 2^m values of the periodogram. The same numbers are produced again if we divide the interval into segments of length 2^{-m}, Walsh transform over each, square, and then average over the segments. It is also worth mentioning that a Walsh transform of 2^n values takes $n2^n$ real additions only; evaluation of S from a by (11) takes $3n2^{n-1}$ real additions and $n2^{n-1}$ divisions by 2. Of course, when we apply a window with cutoff 2^{-m} the computational effort is accordingly reduced. (In other words, n is replaced by $n-m$.)

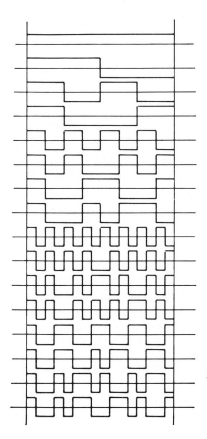

Fig. 12.1

In addition to computational simplicity, Walsh functions also have other nice properties. There is, however, the question of spectrum interpretation. We know of no examples in which pal(i, t) is naturally matched to the physical structures present in $x(t)$. Thus, the utility of Walsh transform as a general tool for analysis is doubtful. There may be cases in which we wish to compute a crude spectrum of some signal for comparison with a standard Walsh spectrum. Here simplicity is paramount and Walsh functions should have the advantage. It is also possible to use Walsh spectra in composition analysis, since in this particular problem a spectrum is simply treated as a function to be mathematically fitted, so that any spectrum, with or without physical significance, would do. (It is also worth mentioning that the 'pilot spectrum' suggested by Blackman and Tukey is a partial Walsh spectrum.)

Finally, let us examine for what kind of signal Walsh transform is optimal. We have by (8)

$$a(s, t) = \Sigma \ \lambda_i \, \text{pal}(i, t) \, \text{pal}(i, s) = \Sigma \ \lambda_i \, \text{pal}(i, t \oplus s) = a(s \oplus t).$$

This is a special type of symmetry which has a simple interpretation if we look at it in connection with an n-dimensional hypercube. It, however, has little resemblance to the physical symmetry of $a(s, t) = a(\,|t-s|\,)$.

Cepstrum

The word is supposed to be pronounced as 'kepstrum', being Tukey's variation of spectrum. Briefly, it is the spectrum of $x'(t) = \log x(t)$. (The assumption is of course that x is positive or at least can be made so without introducing spurious features.) We again give a short discussion to start off the interested reader. There are three basic justifications for the cepstrum:

1. Some signals consist of high peaks interspersed with periods of smaller values. When taking Fourier transform, the contributions from the peaks swamp out those from the smaller and possibly more important values. As the log function increases monotonically but slowly with its argument, the peaks become less important when taking cepstrum and the structures in the smaller values are better analysed.

2. Some signal receptors behave according to a log-arithmic law, i.e., squaring the input doubles the response. Consequently, $\log x(t)$ is of greater physical significance.

3. Direct Fourier transformation of $x(t)$ turns it into a sum of orthogonal functions, whereas physically it may have a multiplicative structure. Fourier transformation of $\log x(t)$ is the same as expressing $x(t)$ as a product of $\exp[y_i(t)]$ values and thus analyses the multiplicative structure.

The cepstrum had its origin in the analysis of seismic signals in which we encounter signals mixed with their own echo, e.g., $x(t) + \alpha x(t + \tau)$. The auto-correlation function is then

$$\langle [x(s) + \alpha x(s+\tau)] \, [x(s+t) + \alpha x(s+t+\tau)] \rangle \sim a(t)(1+\alpha^2) + \alpha a(s-\tau) + \alpha a(s+\tau). \quad (12)$$

And its power spectrum is then approximately

$$(1 + \alpha^2)S_i + 2\alpha S_i \cos(2\pi i \tau) . \tag{13}$$

The echo is normally much smaller than the signal itself, so that $\alpha \ll 1$. The logarithm of (13) gives

$$\sim \ln S_i + 2\alpha \cos(2\pi i \tau) . \tag{14}$$

Now if we take the DFT of (14), we would see a sharp peak at 'frequency' τ. We have thus identified the echo delay, and also its magnitude as the height of the peak increases with α.

In actual seismic signals we get multiple echoes of various amplitudes and time delays, and the echoes do not retain the shape of x exactly. Expression (14) then becomes a sum of cosines with apparently random coefficients and phases. We can then treat it as another random process. Instead of simply Fourier transforming it, we can compute its 'power spectrum' and analyse the echo structure from this.

Appendix 1
Notes on Linear Mathematics

Introduction

This appendix is provided for those readers who occasionally have difficulty with some of the mathematics used in the book. The ideas of linear mathematics are of great generality and are used in numerous areas of practical interest. Familiarity with the material here may make things easier should the reader decide to go on to more advanced reading. The discussion, however, is very incomplete and non-rigorous. There is a lot more to the subject than we can provide here. We do want, however, to advise the reader that: the subject is *not* difficult. What tends to make things hard is the abstract language usually adopted. The reader will find the going easier if he always tries to get an intuitive picture of the mathematical material no matter which book he is reading. We try to do no more than illustrate the type of intuitive picture the reader might get.

We start with the simple case. The relation

$$y = Ax + B,\qquad(1)$$

is linear, because if we plot y against x on graph paper the figure is a straight line. Some common examples are:

(a) The telephone bill, where B is the rental charge, A the charge per call, x the number of calls, and y the total amount payable.

(b) Travel at a constant speed.

(c) Hooke's law, where B is the length of spring when no force is applied, A the elastic constant, x the force applied, and y the length of spring when force is applied.

It should be noted that A and B will vary from example to example. They are adjustable parameters. However, they do not change with x. (Otherwise the relation is no longer linear.) In this sense they are 'constants'.

Equation (1) expresses a single value y in terms of a single value x. These variables are called *scalars*, but quite often, we come into contact with *vectors* and *functions*, in which there may be many different numbers of variables. Further, each value of y may vary with *every* value of x. First consider the case when y is scalar but x is a vector. Then

$$y = \sum_j A_j x_j , \qquad (2)$$

is a linear relation. We can convert (2) into the form of (1) for any particular component of x, say x_k, as

$$y = A_k x_k + B , \qquad \text{where } B = \sum_{j \neq k} A_j x_j .$$

Thus, y is a linear function of each value contained in the vector x. When x is a continuous function of t, then integration replaces addition, and we have, again for a scalar y,

$$y = \int A(t) x(t) \, dt . \qquad (3)$$

Now consider the case when y also contains multiple values, and each value of y is linearly related to each value of x. Different values of y may of course have different coefficients. Thus we have

$$y_i = \sum_j A_{ij} x_j , \qquad (4)$$

i.e., each term y_i is a linear function of each x_j, with coefficient A_{ij}. Similarly for functions of t we have

$$y(t) = \int A(t, t') x(t') \, dt' . \qquad (5)$$

The set of coefficients in (4) form a *matrix*, which is just a common name for a two-dimensional table of numbers. The set of coefficients in (5) form a *kernel*. Both are *linear operators*, because they convert a vector or a function x into another y by means of linear arithmetic.

Frequently, we have physical systems which respond to external stimulus in an approximately linear way. We have already mentioned a spring, force and Hooke's law. Note that, doubling the stimulus (force) doubles the response (increase in length); the response to the sum of two stimuli is the sum of the responses to the individual stimuli. Systems behaving in this way are called linear systems. Up to a point, almost any physical system is approximately linear. Note that (5) satisfies the criterion of linear systems: doubling x doubles y, and the 'response' to 'stimulus' $[x_1(t) + x_2(t)]$ is just

$$\int A(t, t') x_1(t') \, dt' + \int A(t, t') x_2(t') \, dt' ,$$

the sum of the individual responses. In fact, by choosing $A(t, t')$ appropriately we can mathematically approximate the input–output relation of most linear systems.

A particularly important kind of linear system is the linear, time invariant system, for which $A(t, t')$ obeys the restriction

$$A(t, t') = A(t - t') . \qquad (6)$$

Fig. A1 explains the description 'time invariant'. First suppose that the stimulus is a 'shock' at time 0. This is described mathematically by $\delta(t)$, since $\delta(t) = 0$ except at $t = 0$, hence, a shock. The response, $A(t)$, starts at $t = 0$, builds up to some maximum value, and then dies down. Now suppose the shock comes at time s. This is described by $\delta(t-s)$, since $\delta(t-s) = 0$ except if $t = s$, hence a shock at time s. By definition, a time-invariant system would give the same response except that it now starts at $t = s$. It is thus $A(t-s)$. Now if both shocks are applied, the response is the sum of the individual responses because the system is supposed to be linear. Finally, an arbitrary input $x(t)$ can be expressed as a continuous train of shocks at different times. In equation form $x(t) = \int \delta(t-s)\,x(s)\,ds$. (We could look at this differently: $\delta(t-s) = 0$ except at $t = s$, so that it picks out the value $x(t)$ when we integrate.) The response is just the combination of the individual responses to $x(s)\,\delta(t-s)$, or

$$y(t) = \int A(t-s)\,x(s)\,ds. \tag{7}$$

This is just (5) after applying the restriction (6).

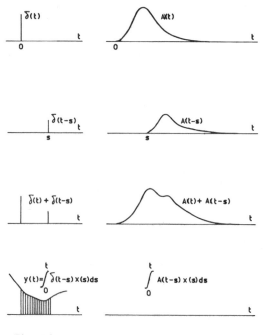

Fig. A1

Equation (7) is a continuous convolution, but not a cyclic convolution. Of course, it is nevertheless possible to compute (7) by means of the Fourier transform if we attach enough zeros to the function $x(t)$.

Real systems are seldom truly linear and time invariant. No system can produce unlimited output if we continuously increase the stimulus. In other words, we get saturation effects. Also, previous stimuli do change the system's response to subsequent input. For example, if the system has already reached saturation, then any later input would produce no additional response at all. Thus, linear, time-invariant systems are no more than idealized, though very useful, models of the real world.

Linear operators

Expressions (4) and (5) are not the only possible types of linear operators. Differentiation, for example, is a linear operation, since the derivative of the sum of two functions is the sum of their derivatives. Because of this, mathematical discussions of linear operators always try to be abstract and general, in order to cover all possible linear operators. While we would not go quite so far, it would be useful to remember that, almost any result we derive for one type of linear operators has an analog for the other types. Depending on which type is more convenient, we shall jump from one to the other during the following discussion. Occasionally, we shall simply write a linear operator as A, and the vector or function it operates on, the *operand*, simply as x. Thus, either (4) or (5) may be written for short as $y = Ax$.

The right-hand sides of (2) and (3) are called *inner products*. They produce scalars from two functions or vectors. To be more exact, the inner product of two vectors x and y is

$$(x, y) = \sum_j x_j^* y_j ,$$

and that between two functions is $\int x^*(t) y(t) \, dt$. When x and y are real the complex conjugation has no effect. We also see that (4) or (5) is just a collection of inner products. It is also clear that, expansion of $x(t)$ as an orthogonal series, say

$$x(t) = \sum_i y_i(t) X_i ,$$

requires evaluation of the inner product of $x(t)$ with each orthonormal function $y_i(t)$: $X_i = \int y_i^*(t) x(t) \, dt$. We can say that each inner product extracts from $x(t)$ the part that is 'similar' to $y_i(t)$, and the combination of all these parts then recovers $x(t)$. The larger X_i, the greater is the portion of $x(t)$ that is 'similar' to $y_i(t)$.

Let us also look at the inner product of a vector or function with itself:

$$(x, x) = \sum_j |x_j|^2 \qquad \text{or} \qquad \int |x(t)|^2 \, dt .$$

This is always real and non-negative. The square root of (x, x) is called the *norm* of x, denoted as $\| x \|$. Clearly, $\| x \|$ is zero only when x_j or $x(t)$ is zero for every j or t. Consequently, the norm indicates how significant a quantity a vector or a function is.

Now let us consider applying two linear operators in succession, say

$$z(t) = \int B(t, t') y(t') \, dt = \iint B(t, t') A(t', t'') x(t'') \, dt'' \, dt' .$$

The overall effect is that of a new operator $C(t, t'')$ with

$$C(t, t'') = \int B(t, t') A(t', t'') \, dt'. \tag{8}$$

(Or, $C = BA$ for short.) The operators B and A each contain a collection of functions used to take inner products with other functions. In (8), however, we are producing a new collection of functions in C by taking inner product of each function in B with each function in A. Similarly, we could produce

$$Z_i = \sum_j C_{ij} x_j \qquad \text{with} \quad C_{ij} = \sum_k B_{ik} A_{kj}.$$

A particular operator, called the *identity operator* and denoted generally as I, leaves its operand unchanged. For (5), the identity operator is none other than $\delta(t - t')$, since its integration with $x(t')$ just picks out its value at t, giving $y(t) = x(t)$. For (4) the identity operator is just δ_{ij}. More concisely, we write $y = Ix = x$. The table of numbers formed by δ_{ij} is called the *unit matrix*. It has 1 on the diagonal, as by definition $\delta_{ii} = 1$ for any i; all the elements off the diagonal are 0, $\delta_{ij} = 0$ for $i \neq j$. Obviously, insertion of an identity operator anywhere in an equation does not change anything. Thus, $IAx = Ax$ and $AIx = Ax$. As result, we can write $IA = A$ and $AI = A$.

For certain operators A we can find B such that applying them in succession leaves the operand unchanged. In other words,

$$\int B(t, t') A(t', t'') \, dt' = \delta(t - t''), \qquad \text{or} \quad BA = I.$$

B is called the *inverse* of A. Note that not every operator has an inverse. Those that have are called *non-singular* operators; the others are called singular operators. We could say that non-singular operators preserve all the information contained in an operand, so that the operand can be recovered. Singular operators destroy part of the information, making recovery impossible. We write the inverse of A as A^{-1}. Note that in addition to $A^{-1} A = I$ we also have $AA^{-1} = I$. This is easily shown by applying A to both sides of the first identity, giving $AA^{-1} A = A$, which shows that AA^{-1} is an operator that leaves A unchanged. In general, however, operator equations cannot be changed in order of precedence, i.e., AB is not necessarily equal to BA. To put it differently, when we have two physical systems it makes a difference whether we send a stimulus into A and then send its response into B, or the reverse.

An operator is *symmetric* if $A(s, t) = A(t, s)$, or $A_{jk} = A_{kj}$. An operator A is *Hermitian* if $A(s, t) = A^*(t, s)$ or $A_{jk} = A^*_{kj}$. If A is real, then being Hermitian it is symmetric also.

Linear operators have wider occurrences than just as models for physical systems. For example, a Fourier transformation [equation (12) in Chapter 2], is a linear operation. It converts a time function $x(t)$ into a vector X_i. Its inverse is just the Fourier series [equation (11) in Chapter 2]. The operator is non-singular as it has an inverse. Another example is the quantity

$ax_1^2 + bx_1 x_2 + cx_2^2$ which can be written as

$$\sum_{i,j=1}^{2} x_i A_{ij} x_j \,, \tag{9}$$

with $A_{11} = a$, $A_{22} = c$ and $A_{12} = A_{21} = b/2$. In other words, it is the inner product between x and Ax. Note that A is a symmetric matrix. The above can easily be generalized to quadratic expressions involving n values of x, in which case (9) would be the inner product of an n-element vector with itself with an $n \times n$ matrix in between.

Eigenvectors

Given a linear operator A, we can always find a set of vectors or functions x^i such that

$$Ax^i = \lambda_i x^i \qquad [\text{or} \quad \int A(t, s) x_i(s) \, ds = \lambda_i x_i(t)] \,. \tag{10}$$

In other words, A operating on x^i reproduces x^i times some constant, rather than a different vector. Of course, when A is the identity operator this holds for any x. When A is something else, it is nevertheless possible to find some x values that satisfy (10). We call x^i the ith *eigenvector* (or sometimes eigenfunction) of A, and λ_i is the ith *eigenvalue*.

An important property of Hermitian (which includes real, symmetric) operators is that eigenvectors corresponding to different eigenvalues are orthogonal. We have

$$\int x_j^*(t)[\lambda_i x_i(t)] \, dt = \int x_j^*(t) A(t, s) x_i(s) \, ds \, dt = [\int x_j(t) A^*(t, s) x_i^*(s) \, ds \, dt]^* \,,$$

where we have moved the conjugation sign outside. (Conjugating anything twice leaves it unchanged.) A is Hermitian, so that $A^*(t, s) = A(s, t)$. We now integrate over t first. Since x_j is an eigenfunction, its integral with $A(s, t)$ produces $\lambda_j x_j(s)$. Thus the above becomes

$$\int x_j^*(t)[\lambda_i x_i(t)] \, dt = [\lambda_j \int x_j(s) x_j^*(s) \, ds]^* = \lambda_j \int x_j^*(s) x_i(s) \, ds.$$

Or

$$(\lambda_i - \lambda_j) \int x_j^*(t) x_i(t) \, dt = 0 \,.$$

Since λ_i and λ_j are different, we have that $\int x_j^*(t) x_i(t) \, dt = 0$.

The eigenvalues of a Hermitian operator are all real. This is shown by deriving $\lambda_i^* = \lambda_i$. We have

$$\int\int x_i^*(t) A(t, s) x_i(s) \, ds \, dt = \lambda_i \int |x_i(t)|^2 \, dt.$$

Take the complex conjugate of both sides. The left-hand side is

$$\int\int x_i(t) A(s, t) x_i^*(s) \, ds \, dt \,,$$

which is the same thing written in a different way. The right-hand side is just $\lambda_i^* \int |x_i(t)|^2 \, dt$. This shows $\lambda_i^* = \lambda_i$, as required. As mentioned earlier, the above two properties also hold for real symmetric operators.

If $x_i(t)$ is an eigenfunction of A, then so is $\alpha x_i(t)$ for any constant α. By choosing α appropriately, we can produce a normalized eigenfunction, i.e., $\int |x_i(t)|^2 \, dt = 1$. Then we have an orthonormal set, and it would be possible to approximate other functions as linear combinations of the eigenfunctions. This gives us an orthogonal series, analogous to the Fourier series. The set may not be complete, however, and there may exist functions which cannot be expressed exactly as such orthogonal series.

Given a set of orthogonal functions, we can in theory form all their possible linear combinations. These together form what we call a *linear space*. Having found all the eigenfunctions of a linear operator and constructed a linear space out of them, we have a neat framework for discussing that particular linear operator. Suppose we have a function from the linear space, say $y(s)$. It can be expressed as

$$y(s) = \sum_j Y_j x_j(s) \qquad \text{or} \qquad y = \sum_j Y_j x^j \, ,$$

for short, Then we have

$$Ay = \sum_j Y_j Ax^j = \sum_j Y_j \lambda_j x^j \, . \qquad (11)$$

This allows us to analyse the operator A in a clear way. In fact, as far as the functions in the linear space is concerned, $A(t, s)$ is indistinguishable from the series

$$\sum_i \lambda_i x_i(t) x_i^*(s) \, , \qquad (12)$$

because operating this on $y(s)$ gives

$$\sum_i \lambda_i x_i(t) \int x_i^*(s) \sum_j Y_j x_j(s) \, ds = \sum_i \lambda_i x_i(t) \sum_j Y_j \delta_{ij} = \sum_i \lambda_i Y_i x_i(t) \, ,$$

which is just (11). Thus, within that linear space the operator can be represented as (12), even though initially the operator may have been defined in some way other than by integration. Expression (12) is called the *spectral representation* of a linear operator.

Let us go a step further. In the linear space each function $y(t)$ is a linear combination of the basic set x_i: $y(t) = \sum Y_i x_i(t)$. Thus, each function in the space corresponds to a *vector* Y. Taking the inner product of two functions, for example, comes to

$$\int z^*(t) y(t) \, dt = \int \sum_i Y_i x_i(t) \sum_j Z_j^* x_j(t) \, dt = \sum_{i,j} Y_i Z_j^* \delta_{ij} = \sum_i Z_i^* Y_i \, ,$$

which is the inner product of these two *vectors* corresponding to y and z. Let us also consider a linear operator B which converts function $y(t)$ into another function, say $z(t)$. Both can be expanded in terms of the basic functions x. The eigenfunctions of B are of course different from the x, which are eigenfunctions of A. Let us say that B operating on $x_i(t)$ converts it into $b_i(t)$, which can again be expressed as a linear combination of the x functions:

$$b_i(t) = \sum_j b_j^i x_j(t) \, .$$

(The superscript i indicates that each $b_i(t)$ is different and so has different coefficients.) Now we have

$$z = By = \sum_i Y_i Bx^i = \sum_{i,j} Y_i b^i_j x^j .$$

Since $z = Z_j x^j$, we have

$$Z_j = \sum_i b^i_j Y_i .$$

This is the multiplication of a *matrix* into the vector representing y to produce a vector representing z. Thus, the matrix formed by the value b^i_j represent the operator B, even though it was defined first in a non-matrix way.

And here lies the source of the power of linear mathematics. Every linear operator has a corresponding matrix which behaves like the operator in all the essential mathematical relations; two linear operators corresponding to the same matrix would behave in the same way, even though they have been defined in very different forms. All kinds of linear operators are in fact equivalent. This is what makes linear mathematics so general and elegant, and why the language used to study it is usually so abstract.

Orthogonalization

Given a set of M non-orthogonal functions or vectors, there is a simple procedure by which we may produce a set of orthogonal ones from them. The process, however, is not guaranteed to produce M orthogonal functions. Whether it does depends on whether the given functions are *linearly independent*. If they are not, then some of the functions we try to produce would turn out to be identically zero.

First let us take only two functions, $x_1(t)$ and $x_2(t)$. Clearly, the function

$$\hat{x}_2(t) = x_2(t) - x_1(t)[\int x^*_1(t) x_2(t) \, dt / \int |x_1(t)|^2 \, dt] , \tag{13}$$

is orthogonal to $x_1(t)$, as integrating it with $x^*_1(t)$ gives zero. In a sense, we have subtracted from $x_2(t)$ the part that is 'similar' to $x_1(t)$, leaving only the part that is 'different' or 'independent'. Note that if x^1 and x^2 happen to be equal, (13) is identically zero. Or, if x^2 has no part that is independent of x^1 then we cannot orthogonalize them.

Now let us generalize the above process to M functions $x^1 \ldots x^M$. We with to produce, if possible, M *orthonormal* functions $y1 \ldots yM$. We first produce

$$y_1(t) = x_1(t) / \| x^1 \| ,$$

where $\| x \|$ is the norm of x defined on page 137. That $y_1(t)$ is indeed normalized can be verified readily by integrating its square. We then produce $y_2(t)$ by

$$y_2(t) = \hat{x}_2(t) / \| \hat{x}^2 \| ,$$

with

$$\hat{x}_2(t) = x_2(t) - y_1(t)(y^1, x^2) .$$

We could write

$$y_2(t) = x_2(t)/\|\hat{x}^2\| - x_1(t)(y^1, x^2)/(\|x^1\|\|\hat{x}^2\|). \tag{14}$$

Similarly we have $\hat{x}_3(t) = x_3(t) - y_2(t)(y^2, x^3) - y_1(t)(y^1, x^3)$, which is again normalized to produce $y_3(t)$. In general we have

$$\hat{x}_i(t) = x_i(t) - \sum_{j=1}^{i-1} y_t(t) \int y_j^*(t) x_i(t)\, dt, \tag{15}$$

and

$$y_i(t) = \hat{x}_i(t)/\|\hat{x}^i\|. \tag{16}$$

Recursively, we produce $y_i(t)$ for larger i until we reach M.

During the process we may find, however, that some of the x^i produce 0 as $\hat{x}^i$. In other words, x^i may have no part that is independent of the previously produced functions $y^1 \dots y^{i-1}$. Consequently, we cannot produce M orthogonal functions. We describe this by saying that the given functions x are linearly dependent, since x^i can be expressed exactly as a linear combination of y functions of smaller indices, which means x^i can also be expressed as a sum of x functions because the y is produced from the x. Since we do not have a $y_i(t)$, when we go on to produce $y_{i+1}(t)$, $y_{i+2}(t)$, etc., the summation in (15) would go from 1 to $(i-1)$, and then skip i for larger indices.

In actual computation we almost never get exact zeros. Rather, we would obtain an $\hat{x}^i$ which is very small for most values of t. Thus, $\|\hat{x}^i\|$ is very small, and we have to greatly enlarge $\hat{x}^i$ in order to produce y^i by (16). This magnifies any errors $\hat{x}^i$ might have. Since y^i is made unreliable, all subsequent y functions are also unreliable.

We often have to solve problems of the following form. Find X such that the following is satisfied as closely as possible:

$$\sum_i X_i x_i(t) \sim f(t). \tag{17}$$

The obvious method was that shown on page 6, where we minimized the mean-square error and came up with the M equations for X:

$$\sum_j X_j \int x_i(t) x_j^*(t)\, dt = \int x_j^*(t) f(t)\, dt. \tag{18}$$

We saw in Chapter 10 (page 111) that similar equations are used for composition analysis. As we said then, if the x are close to being linearly dependent, then the solution of (18) is highly unreliable, just as the orthogonal set produced from the x functions would be unreliable.

However, some recent studies of numerical analysts show that, instead of simply solving (18) we should first orthogonalize the x to obtain the y, then integrate both sides of (17) with $y_j(t)$ to produce

$$\sum_i X_i \alpha_{ij} = F_j, \tag{19}$$

where

$$\alpha_{ij} = \int x_i(t) y_j(t)\, dt, \tag{20}$$

and
$$F_j = \int f(t)\,y_j(t)\,dt\,, \tag{21}$$

and finally solve (19) for the X. It was shown that the results produced in this way by what we call the orthogonalization procedure are far more reliable. Note that (20) implies $x_i(t) = \sum \alpha_{ij}\,y_j(t)$, as (20) just defines the orthogonal transform of $x_i(t)$.

In actual computation, it is not necessary to either compute α_{ij}, or solve (19) at all. Instead, we require the matrix β that satisfies

$$y_k(t) = \sum_i \beta_{ki}\,x_i(t)\,. \tag{22}$$

Since y functions have been produced from the x, this matrix is already available. For example, we had $y_1(t) = x_1(t)/\|x^1\|$, so that $\beta_{11} = \|x^1\|^{-1}$ and $\beta_{1j} = 0$ for any other j. By (14) we have $\beta_{22} = \|x^2\|^{-1}$ and $\beta_{21} = (y^1, x^2)/(\|x^1\|\,\|x^2\|)$ with $\beta_{2j} = 0$ for other j. In short we can compute β as we are orthogonalizing the x. Then, we produce X by

$$X_i = \sum_j \beta_{ji}\,F_j\,. \tag{23}$$

This does indeed satisfy the original requirement, (17), as

$$\sum_i X_i\,x_i(t) = \sum_{ij} \beta_{ji}\,x_i(t)F_j = \sum_j y_j(t)F_j\,,$$

where we have used (22). The above is none other than the orthogonal expansion of $f(t)$ in terms of the y functions, which is a least-square error approximation. Thus, the X_i found from (23) do satisfy (17).

In short, in the orthogonalization procedure we first produce the y functions from the x, taking care to preserve the coefficients β that express y in terms of x. Then we integrate y with f to obtain F, and finally use (23) to produce X.

The set of powers, t^i, $i = 0, 1, ..., M$, is often used in function approximation. We saw in Chapter 4 their use in trend removal. The set is linearly independent, but when M becomes large it approaches dependence. Consequently, the usual method for polynomial fitting, by the use of (18), is very unsuitable for M over 6 or 7. (Unfortunately, many computer users do not realize this, and waste much time trying to fit high-order polynomials to data this way, producing nearly useless results.) If we orthogonalize the functions over the integration interval $[-1, 1]$, the orthogonal functions we obtain are the Legendre polynomials. Books on partial differential equations usually list these polynomials up to some maximum order, so that the matrix β, which relates y_i (Legendre) to x_i (t^i), can be got just by looking them up. [Note, however, that the books generally give the polynomials in unnormalized form. In other words, they are really $\hat{x}_i(t)$ rather than $y_i(t)$.] Orthogonalization over different intervals would produce other kinds of polynomials. Thus, to make use of Legendre polynomials we should scale the interval over which our data have been measured into $[-1, 1]$. This standardization is also desirable from the reliability point of view, as the t^i functions are less dependent over $[-1, 1]$ than other intervals.

By taking all possible linear combinations of the set of orthogonal functions which we have produced, we again obtain a linear space. Now, since the y functions are obtained from the x, a linear combination of the y functions is also a linear combination of the x. If the x functions are linearly independent, then there is a unique representation for any function in the linear space in terms of these x. Otherwise the linear combination is not unique. In the former case, we say that the x functions form a *basis* of the linear space, in the sense that they can be used to produce the linear space. A linear space can have many different bases. For example, the y are obviously a basis, as a function within the linear space certainly has a unique representation in terms of them.

In (17) the left-hand side is a function within the linear space, since it is a linear combination of the x functions. The right-hand side, being arbitrary, may be a function outside the space. The problem posed by (17) is therefore to find a function from within the space that best approximates $f(t)$. As we said, we could find this by solving (18), but it is best to proceed via an orthogonal basis set, the y. This type of 'best approximation' problem occurs in many applications. Filtering, for example, is one such problem. The filter we construct is always limited in being able to produce certain kinds of functions but not others, and our task is to choose an output that best reproduces the useful information in the input out of the possible set of outputs. Problems of forecasting are similar. We want to take certain combinations of the past values that best predict the future. In view of the importance of such problems, and because of the clear relevance of the theory of linear space to such problems, it does the reader no harm to learn a little more about the subject. We hope to have put him on the right track with this introduction.

Appendix 2
Computer Programs

This appendix lists five Fortran subroutines designed to perform some of the operations most frequently used in spectral analysis. In writing, we have chosen to keep them simple at the expense of efficiency in order that the reader can understand them easily. As long as they are used for problems within the limits we prescribed, there is no excessive wastage of time and storage. For those who wish to start experimenting with spectral analysis techniques the subroutines should make things very convenient. Once the reader gets into serious data analysis he would want to write his own, more efficient, and more specialized computer programs. Even then, the availability of these subroutines should facilitate the reader's programming effort.

No detailed explanation of these subroutines will be given here as each has its own comments. A short list is given below:

1. FFT: for both forward and inverse transform of *complex* vectors.

2. FFTR: for the forward transform of a *real* vector or its recovery from its DFT. (X_i for $i = 0, 1, ..., \frac{1}{2}N$ only.)

3. PERIOD: computes the periodogram of a real vector at half integer frequencies, $i/2$, $i = 0, 1, ..., N-1$.

4. AUTCOR: computes the autocorrelation estimate of N given values up to time delay M.

5. COTRAN: computes the Fourier transform of a real, even vector, also known as a cosine transform. It returns the power spectrum if given the autocorrelation function.

```
        SUBROUTINE FFT(A,M,IS)
C FFT OF COMPLEX ARRAY A, OF 2**M ELEMENTS, IS=+1 OR -1 SIGN OF CEXP
C (NOTE THAT INITIAL DATA IN ARRAY A IS REPLACED BY ITS FOURIER TRANSFORM)
C  FIRST PART IS BIT-REVERSED PERMUTATION USING RECURSIVE ALGORITHM,
C  WHICH INCREMENTS A REVERSED INDEX WHEN NEEDED FOR EACH BIT POSITION
C FINAL PART, FROM LABEL 7, IS BASE-2 FFT COMPUTATION,
C  WHICH REQUIRES MINIMUM DIFFERENT W, GENERATED RECURSIVELY
        COMPLEX A(1),TEMP,W,D
        INTEGER IRA(16),NR(16),SPAN,STEP
        DATA PI/3.141592653589793/
        N=2**M
        DO 1 J=1,M
        IRA(J)=0
1       NR(J)=2**(J-1)
C REVERSED INDEX SETS (FOR EACH BIT POSITION) INITIALISED
        IF=1
2       IR=IRA(M)+1
        IF(IR.LE.IF)GO TO 3
C PREVENTS NULLIFYING DOUBLE SWAP
        TEMP=A(IF)
        A(IF)=A(IR)
        A(IR)=TEMP
C REVERSED INDEX PAIR SWAPPED
3       IF=IF+1
C INCREMENT FORWARD INDEX IF
        IF(IF.GT.N)GO TO 7
        J=M
4       IF(IRA(J).LT.NR(J))GO TO 5
C ALTERNATE INCREMENT OF IRA(J), MUST GO BACK ONE BIT
        J=J-1
        GO TO 4
5       IRA(J)=IRA(J)+NR(J)
C SIMPLE, ALTERNATE INCREMENT OF REVERSED INDEX
6       IF(J.EQ.M)GO TO 2
        IRA(J+1)=IRA(J)
C WORK FORWARD THROUGH REVERSED INDEX BIT SET
        J=J+1
        GO TO 6
C ARRAY IS NOW IN BIT-REVERSED ORDER, M COMPUTING PASSES FOLLOW
7       DO 9 J1=1,M
        SPAN=2**(J1-1)
        STEP=2*SPAN
C SPAN BETWEEN ELEMENTS IN PAIR, STEP TO NEXT PAIR WITH SAME W
        W=(1.,0.)
        D=CEXP(CMPLX(0.,PI/SPAN))
        IF(IS.LT.0)D=CONJG(D)
C STARTING PHASE ADJUSTER W, MODIFIER D
        DO 9 J2=1,SPAN
        DO 8 J=J2,N,STEP
        K=J+SPAN
        TEMP=A(K)*W
        A(K)=A(J)-TEMP
8       A(J)=A(J)+TEMP
C INNER LOOP ARITHMETIC - TWO POINT TRANSFORMS
9       W=W*D
C RECURSIVE MODIFICATION OF PHASE ADJUSTER W
        RETURN
        END
```

```
      SUBROUTINE FFTR(A,M,IS)
C REAL-TO-COMPLEX (OR VICE-VERSA) HALF-LENGTH FFT OF ARRAY A
C (NOTE THAT INITIAL DATA IN ARRAY A IS REPLACED BY ITS FOURIER TRANSFORM)
C  REAL DATA ASSUMED PACKED ALTERNATELY AS REAL AND IMAGINARY VALUES,
C  MOST EASILY ACHIEVED BY EQUIVALENCING REAL AND COMPLEX ARRAY NAMES
C  2**M REAL ELEMENTS (+2 DUMMIES) OR 2**(M-1)+1 COMPLEX ELEMENTS
C  IS=+1 OR -1 SIGN OF CEXP AND DIRECTION (+1=REAL-TO-COMPLEX, -1 REVERSE)
C USES SCRAMBLE/UNSCRAMBLE ALGORITHM AND CALL TO HALF-LENGTH COMPLEX FFT
      COMPLEX A(1),TA,TB,W,D
      DATA PI/3.141592653589793/
      MH=M-1
      N=2**MH
      INCNT=N/2+1
      W=(1.,0.)
      D=CEXP(CMPLX(0.,PI/N))
C STARTING PHASE ADJUSTER W, MODIFIER D FOR SCRAMBLE/UNSCRAMBLE
      IF(IS.LT.0)GO TO 2
C REAL-TO-COMPLEX FFT FOLLOWS, HALF-LENGTH COMPLEX FFT FIRST
      CALL FFT(A,MH,IS)
      A(N+1)=A(1)
      DO 1 J=1,INCNT
      K=N+2-J
      TA=(A(J)+CONJG(A(K)))*0.5
      TB=-CONJG(A(J))+A(K)
      TB=CMPLX(AIMAG(TB),REAL(TB))*W*0.5
      A(J)=TA+TB
      A(K)=CONJG(TA-TB)
1     W=W*D
C ELEMENTS UNSCRAMBLED, W RECURSIVELY MODIFIED
      RETURN
C COMPLEX-TO-REAL FFT FOLLOWS
2     D=CONJG(D)
      DO 3 J=1,INCNT
      K=N+2-J
      TA=A(J)+CONJG(A(K))
      TB=(A(J)-CONJG(A(K)))*W
      TB=CMPLX(AIMAG(TB),REAL(TB))
      A(J)=TA-CONJG(TB)
      A(K)=CONJG(TA)+TB
3     W=W*D
C ELEMENTS SCRAMBLED, W RECURSIVELY MODIFIED
      CALL FFT(A,MH,IS)
C HALF-LENGTH COMPLEX FFT FINISHES COMPLEX-TO-REAL FFT
      RETURN
      END
```

```
      SUBROUTINE PERIOD(N,DATA,PDGRAM)
C     THIS SUBROUTINE ACCEPTS  N  INPUT VALUES AND RETURNS THEIR PERIODOGRAM.
C     N MUST NOT EXCEED 512.  THE METHOD IS BAD FOR LARGE  N.
      DIMENSION DATA(N),PDGRAM(N),FIXCOS(513),FIXSIN(513)
      DATA NSAVE/0/
      NN=N+1
      N2=N*2
      NN2=NN*2
      REC=1./FLOAT(N)
C     THE LOOP BELOW STORES VALUES OF SINE AND COSINE BETWEEN  0  AND  PI.
C     IF  NSAVE=N  THEN THE SUBROUTINE HAS BEEN CALLED EARLIER WITH THE SAME  N
C     AND SO MUST ALREADY CONTAIN CORRECT FIXCOS AND FIXSIN.
      IF(NSAVE.EQ.N)GO TO 10
      REC2=REC*4.*ATAN(1.)
C     THIS IS  PI/N.
      DO 5 I=1,NN
      ARG=FLOAT(I-1)*REC2
      FIXCOS(I)=COS(ARG)
      FIXSIN(I)=SIN(ARG)
    5 CONTINUE
   10 CONTINUE
      REC=REC*REC
      DO 20 I=1,N
C     TEMP1 AND TEMP2 WILL BE THE REAL AND IMAGINARY PARTS OF THE  DFT  OF DATA
      TEMP1=DATA(1)
      TEMP2=TEMP1
      II=I-1
C     K  IS THE VALUE OF  I*J  ATER SUBTRACTION OF MULTIPLES OF  2N.
      K=1
      DO 15 J=2,N
      K=K+II
      IF(K.GT.N2)K=K-N2
      IF(K.GT.NN)GO TO 12
C     ARGUMENT OF SINE AND COSINE NOT OVER  PI.
      A=FIXCOS(K)
      B=FIXSIN(K)
      GO TO 13
C     ARGUMENT OF SINE AND COSINE MORE THAN  PI.  USE  SIN(ARG)=-SIN(2*PI-ARG),
C     COS(ARG)=COS(2*PI-ARG).
   12 KK=NN2-K
      A=FIXCOS(KK)
      B=-FIXSIN(KK)
   13 D=DATA(J)
      TEMP1=TEMP1+D*A
      TEMP2=TEMP2+D*B
   15 CONTINUE
C     SQUARE REAL AND IMAGINARY PARTS AND ADD TO GIVE POWER.
      PDGRAM(I)=(TEMP1*TEMP1+TEMP2*TEMP2)*REC
   20 CONTINUE
      NSAVE=N
      RETURN
      END
```

```
      SUBROUTINE AUTCOR(N,M,DATA,COR)
C   N  IS THE NUMBER OF INPUT DATA,   M  THE NUMBER OF AUTOCORRELATIONS NEEDED
C   M  SHOULD NOT BE MORE THAN   256.   THE METHOD IS BAD FOR LARGE  M.
      DIMENSION DATA(N),COR(M)
      REC=1./FLOAT(N)
      DO 10 I=1,M
      TEMP=0.
      DO 5 J=I,N
      JJ=J -I+1
      TEMP=TEMP+DATA(J)*DATA(JJ)
    5 CONTINUE
      COR(I)=TEMP*REC
   10 CONTINUE
      RETURN
      END



      SUBROUTINE COTRAN(M,COR,SPECTR)
C   THIS SUBROUTINE ACCEPTS  M  AUTOCORRELATION VALUES AND RETURNS THE REAL
C   PARTS OF THEIR FOURIER TRANSFORM, I.E., UNWINDOWED SPECTRUM. WINDOWING
C   MAY BE APPLIED EITHER BY MULTIPLICATION BEFORE CALLING THIS SUBROUTINE,
C   OR BY AVERAGING NEIGHBOURING TERMS AFTER RETURN.
C   M  MUST NOT EXCEED 128.   THE METHOD IS BAD FOR LARGE  M.
      DIMENSION COR(M),SPECTR(M),FIXCOS(129)
      DATA MSAVE/0/
      REC=1./FLOAT(M)
      MM=M+1
      M2=M*2
      MM2=MM*2
C   THE LOOP BELOW STORES THE VALUES OF COSINE BETWEEN   0  AND  PI.
C   IF MSAVE IS 0 THE SUBROUTINE HAS NOT BEEN CALLED BEFORE. IF IT IS EQUAL
C   TO   M   THEN FIXCOS ALREADY CONTAIN CORRECT VALUES.
      IF(NSAVE.EQ.M)GO TO 10
      REC2=REC*4.*ATAN(1.)
      DO 5 I=1,MM
      ARG=FLOAT(I-1)*REC2
      FIXCOS(I)=COS(ARG)
    5 CONTINUE
   10 HALF=COR(1)*0.5
      REC=REC*2.
      DO 20 I=1,M
      TEMP=HALF
      II=I-1
C   K  IS THE VALUE OF  I*J  REDUCED BY MULTIPLES OF  2M.
      K=1
      DO 15 J=2,M
      K=K+II
      IF(K.GT.M2)K=K-M2
      KK=K
C   K   GREATER THAN  M+1  MEANS ARGUMENT OF COSINE IS MORE THAN  PI.  USE
C   COS(ARG)=COS(2*PI-ARG).
      IF(KK.GT.MM)KK=MM2-KK
      TEMP=TEMP+COR(J)*FIXCOS(KK)
   15 CONTINUE
      SPECTR(I)=TEMP*REC
   20 CONTINUE
      MSAVE=M
      RETURN
      END
```

Further Reading

The following list includes books on spectral analysis as well as closely related topics. Commendations and cautions are given where appropriate. We give no list of papers in this book, as any reasonably complete list would be very long, and in any case it would be of marginal benefit to those other than spectral analysis specialists. However, several of the listed books contain extensive bibliographical material, particularly [2] [3].

Spectral analysis

[1] G. M. Jenkins and D. G. Watts, *Spectral Analysis and Its Applications*, Holden-Day, San Francisco, 1968.

(Ten years old, but still the best for learning about the subject. Excellent introduction to theoretical background.)

[2] L. H. Koopmans, *The Spectral Analysis of Time Series*, Academic Press, New York, 1974.

[3] D. R. Brillinger, *Time Series: Data Analysis and Theory*, Holt, Rinehart and Winston, 1975.

[4] P. Bloomfield, *Fourier Analysis of Time Series: An Introduction*, Wiley, New York, 1976.

(Three statistics oriented texts. The last is rather brief.)

[5] J. S. Bendat and A. G. Piersol, *Random Data: Analysis and Measurement Procedures*, Wiley, New York, 1971.

[6] R. K. Otnes and L. Enochson, *Digital Time Series Analysis*, Wiley, New York, 1972.

(Two engineering oriented books.)

[7] J. N. Rayner, *An Introduction to Spectral Analysis*, Pion Limited, London, 1971.

[8] R. W. Harris and T. J. Ledwidge, *Introduction to Noise Analysis*, Pion Limited, London, 1974.

[9] R. B. Blackman and J. W. Tukey, *The Measurement of Power Spectra*, Dover, New York, 1959.

[10] B. Harris, *Spectral Analysis of Time Series*, Wiley, New York, 1967.

Theory of time series

[11] G. E. Box and G. M. Jenkins, *Time Series Analysis: Forecasting and Control*, Holden-Day, San Francisco, 1971.

[12] E. J. Hannan, *Time Series Analysis*, Methuen, London, 1960.

[13] E. J. Hannan, *Multiple Time Series*, Wiley, New York, 1970.

[14] E. Parzen, *Time Series Analysis Papers*, Holden-Day, San Francisco, 1968.

[15] M. G. Kendal, *Time-Series*, Griffin, London, 1973.

(The above five are only a small selection of the field.)

Digital signal processing

[16] L. R. Rabiner and B. Gold, *Theory and Application of Digital Signal Processing*, Prentice-Hall, Englewood Cliffs, N.J., 1975.

(The best on the subject.)

[17] A. V. Oppenheim and R. W. Schafer, *Digital Signal Processing*, Prentice-Hall, Englewood Cliffs, N.J., 1975.

(The best undergraduate text on the subject.)

[18] B. Gold and C. M. Rader, *Digital Processing of Signals*, McGraw-Hill, New York, 1969.

[20] M. Schwartz and I. Shaw, *Signal Processing: Discrete Spectral Analysis, Detection and Estimation*, McGraw-Hill, New York, 1975.

[21] W. D. Stanley, *Digital Signal Processing*, Reston, New York, 1975.

[22] S. D. Stearns, *Digital Signal Analysis*, Haydon Book Co., Rochelle Park, N.Y., 1975.

[23] R. E. Bogner and A. G. Constantinides, *Introduction to Digital Filtering*, Wiley, New York, 1975.

[24] K. G. Beauchamp, *Signal Processing*, George Allen & Unwin, London, 1973.

[25] M. H. Ackroyd, *Digital Filtering*, Butterworths, London, 1973.

(The set of books cover roughly the same topics as those on spectral analysis, but with different emphasis. They are all engineering oriented, and assume background in engineering mathematics, with the exception of [24], which is relatively unmathematical, but unfortunately contains a fair number of inaccuracies.)

Fourier series

[26] E. O. Brigham, *The Fast Fourier Transform*, Prentice-Hall, Englewood Cliffs, N.J., 1974.

[27] N. Ahmed and K. R. Rao, *Orthogonal Transforms for Digital Signal Processing*, Springer, Berlin/New York, 1975.

[28] R. Bracewell, *The Fourier Transform and Its Applications*, McGraw-Hill, New York, 1965.

[29] C. Lanczos, *Discourse on Fourier Series*, Oliver and Boyd, Edinburgh, 1966.

(Again a limited selection. [28] is a popular text. [29] is a very readable mathematics book, which is quite rare. Most recent texts on spectral analysis and digital signal processing include extensive discussion of the fast Fourier transform.)

Miscellaneous

[30] D. Newland, *Random Vibrations and Spectral Analysis*, Longman, London, 1975.

[31] M. Bath, *Spectral Analysis in Geophysics*, Elsevier, Amsterdam, 1974.

[32] G. S. Fishman, *Spectral Methods in Econometrics*, Harvard University Press, 1969.

[33] J. A. Blackburn, *Spectrum Analysis*, Dekker, New York, 1970.

[34] R. J. Bell, *Introductory Fourier Transform Spectroscopy*, Academic Press, New York, 1972.

[35] W. R. Davenport and W. L. Root, *An Introduction to Random Signals and Noise*, McGraw-Hill, New York, 1958.

[36] J. B. Thomas, *An Introduction to Statistical Communication Theory*, Wiley, New York, 1969.

Index

Digital Spectral Analysis

C. K. Yuen
Department of Information Science
University of Tasmania

D. Fraser
Division of Computing Research
CSIRO, Canberra

Digital Spectral Analysis is a self-contained introduction to
spectral analysis and digital signal processing that requires
no prior knowledge of engineering mathematics, communication
theory or advanced probability theory. Any scientist or
engineer with a basic working knowledge of calculus,
elementary statistics and matrix algebra will be able to gain
a complete appreciation of the power and utility of the most
recent developments in the mathematical techniques of
spectral analysis.

ISBN 0 643 02419 0   Recommended retail price $8.50